# A-LEVEL GEOGRAPHY TOPIC MASTER

# GLOBAL SYSTEMS

Series editor
Simon Oakes

Simon Oakes

**HODDER**
EDUCATION
AN HACHETTE UK COMPANY

The Publishers would like to thank the following for permission to reproduce copyright material.

**Dedication**

For Victoria T, Nick L and Chris B – you made and supported the curriculum and classes from which this book grew.
For the Bancroft's geographers who came to those classes – this book is all the better for your many ideas and questions.

I am grateful to Ruth Murphy at Hodder for developing the Geography Topic Master series and inviting me to write this title. All books in the series have benefited from the expertise and support provided by Margaret McGuire, Rob Bircher and Lydia Hardman. I am indebted too to Noel Castree, Professor of Geography at Manchester University, who acted as academic reviser for this book. Finally, thank you to Rosalind Oakes for helping me prepare the manuscript and providing encouragement at moments when it was most needed.

**Photo credits and Acknowledgements can be found on page 220**

Every effort has been made to trace all copyright holders, but if any have been inadvertently overlooked, the Publishers will be pleased to make the necessary arrangements at the first opportunity.

Orders: please contact Bookpoint Ltd, 130 Park Drive, Milton Park, Abingdon, Oxon OX14 4SE. Telephone: +44 (0)1235 827827. Fax: +44 (0)1235 400401. Email education@bookpoint.co.uk Lines are open from 9 a.m. to 5 p.m., Monday to Saturday, with a 24-hour message answering service. You can also order through our website: www.hoddereducation.co.uk

ISBN: 978 1 5104 2793 8

© Simon Oakes 2019

First published in 2019 by
Hodder Education,
An Hachette UK Company
Carmelite House
50 Victoria Embankment
London EC4Y 0DZ
www.hoddereducation.co.uk

Impression number   10 9 8 7 6 5 4 3 2 1

Year                 2023  2022  2021  2020  2019

Cover photo © Paul White - Transport Infrastructures / Alamy Stock Photo
Illustrations by Aptara Inc.
Typeset in India by Aptara Inc.
Printed in Slovenia by DZS GRAFIK D.O.O.
A catalogue record for this title is available from the British Library.

# Contents

# Introduction

This book's review of contemporary global systems demonstrates how much *geography really matters*. Just think of recent events such as: the 2016 Brexit referendum; trade rivalry between the USA, China and other powers; climate change; and the reshaping of how we live by new technologies created by Google and other powerful transnational (or multinational) corporations. The geographical concepts around which this book is organised are now defining political ideas of our time. They include globalisation and its power relations (Chapters 1 and 2), the pros and cons of global interdependence (Chapter 3), global inequality (Chapter 4) and the subsequent injustices suffered by local communities (Global 5). Chapter 6 explores the relationship between these issues and a recent surge in popular opposition against global systems.

## The A-level Geography Topic Master series

The books in this series are designed to support learners who aspire to reach the highest grades. To do so requires more than learning by rote. Only around one-third of available marks in an A-level Geography examination are allocated to the recall of knowledge (*assessment objective 1, or AO1*). A greater proportion is reserved for higher-order cognitive tasks, including the **analysis**, **interpretation** and **evaluation** of geographic ideas and information (*assessment objective 2, or AO2*). The material in this book encourages active reading and critical thinking. The overarching aim is to help you develop the analytical and evaluative 'geo-capabilities' needed for examination success. Opportunities to practise and develop **data manipulation skills** are also embedded throughout the text (supporting *assessment objective 3, or AO3*).

All *Geography Topic Master* books prompt students constantly to 'think geographically'. In practice this can mean learning how to seamlessly integrate **geographical concepts** – including globalisation, causality, place, scale, identity, inequality, interdependence, feedback and risk – into the way we think, argue and write. The books also take every opportunity to use page-referencing to establish **synoptic links** (this means making 'bridging' connections between themes and topics). Additionally, connections have been highlighted between *Global systems* and other Geography topics such as *Changing places* or *Water and carbon cycles*.

## Using this book

The book may be read from cover to cover since there is a logical progression between chapters. On the other hand, a chapter may be read independently whenever required as part of your school's scheme of work for this topic. A common set of features are used in each chapter:

- *Aims* establish the main points (and sections) of each chapter.
- *Key concepts* are important ideas relating either to the discipline of Geography as a whole or more specifically to the study of global systems.
- *Contemporary case studies* apply geographical ideas, theories and concepts to real-world contexts.
- *Analysis and interpretation* features help you develop the geographic skills needed for the application of knowledge and understanding (AO2) and data manipulation (AO3).
- *Evaluating the issue* brings each chapter to a close by discussing a key global systems issue (typically involving competing perspectives and views).
- At the end of each chapter are the *Chapter summary, Refresher questions, Discussion activities, Fieldwork focus* (supporting the independent investigation) and selected *Further reading.*

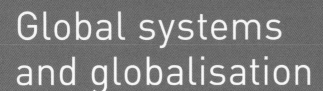

**CHAPTER 1**

# Global systems and globalisation

A step change in the complexity of global systems has occurred in recent decades. People, places and environments are far more densely interconnected and interdependent now than they used to be. This chapter:

- analyses the linked concepts of global systems and globalisation
- investigates how transport and communications technology help give rise to the trade, migration, money and information flows which link places together
- explores the importance of trade and transnational corporations (TNCs) for global systems growth
- discusses the interrelationships between technology, TNCs and globalisation.

## KEY CONCEPTS

**Global systems (or networks)** The global-scale economic, social and political structures that are created when human beings interact with one another across national borders at planetary and world-region scales. Flows of money, people, merchandise, services and ideas link together people, places and environments.

**Interdependence** Relations of mutual dependence between humans and/or non-human things. For example, states may become dependent on one another's human and physical resources as a result of trade and migration flows.

**Time–space compression** Heightened connectivity changes our perception of time, distance and potential barriers to the movement of people, goods, money and information. As travel and communications times fall due to new inventions, different places approach each other in 'space-time': they feel closer together than in the past. This idea is central to geographer David Harvey's work.

**Causality** The relationship between cause and effect. Everything has a cause or causes. For example, migration is triggered by push and pull factors, along with the technology that facilitates movement.

# Global systems, flows and processes

▶ *How are local places linked together by different global systems and flows?*

## Global systems theory

The idea of global systems is used by human geographers in order to understand how human activity operates on a worldwide scale. In physical geography, system theory is widely used, for example in the study of landscapes, ecosystems or meteorology. Physical systems are conceptualised as open or closed structures whose different parts are linked together by flows of energy and matter.

A similar approach may be taken in human geography. At a local scale, settlements or particular clusters of industries can be thought of as systems with inputs and outputs of materials, information, products and people. Studies of human activity at a planetary scale are based on four main principles:

1  All of the world's people, industries and national economies form parts of a single unified, complex and interconnected structure. Even farmers in the world's remotest rural areas will most likely be connected with other distant people and places in some way, whether it is through sales of their surplus crops, receipt of international aid or via text messages exchanged with family members who have migrated to cities or other countries.

2  The connections that link people and places together can be viewed as 'flows'. In the same way that physical systems are driven by flows of energy and matter, human systems depend on flows of money, people, ideas and commodities (in the sense that these different flows are the circulating 'blood' which helps nourish the 'body' of global systems). Sometimes these flows maintain balance and stability for different societies; at other times they may operate in ways that disrupt equilibrium and bring negative change. New ideas and technologies, as they flow from place to place and are adopted by different societies, may radically alter the way people live and think on a worldwide scale. In today's **hyperconnected** world, the connections of money, information and materials that link businesses together often span continents as part of a single global economic system.

3  All global flows – and the players they connect together – are embedded in a broad political and legal framework of rules and conventions. Over time, a near-universal consensus has formed that views such things as free trade and property rights as **global norms**. Local economic activities that generate goods and services are part of this greater politico-economic structure. Alongside this framework, there exist widely (though not completely) accepted social norms and conventions which help to regulate human activity at both global and local scales. A good example of this is the United Nations' work promoting and protecting the universal principle of human rights.

4  The complexity of global systems has heightened human interconnectivity in ways that have led in turn to far greater interdependence. This means that people, places, businesses and countries have become mutually dependent on one another for their continued prosperity and wellbeing: few societies today could be described as being entirely self-sufficient. Interdependent relationships are the focus of Chapter 3. Deep interdependency is today far more common than weak interdependency.

## 🔑 KEY TERMS

**Hyperconnected** A state which exists when the connections in a system have increased to the point where the linkages between system elements (people and places) have become numerous and dense.

**Global norms** Shared standards of acceptable behaviour for the world's sovereign state governments. This definition of global norms encompasses almost any area of national rule-making, ranging from environmental and wildlife protection issues to economic and cultural matters (such as import taxes or equality for minorities, respectively).

The growth of global systems over time has transformed the economic life of local places, in turn also bringing social and cultural transformations. These processes of change are collectively called globalisation. Figure 1.1 shows how the concepts of globalisation and global systems are related.

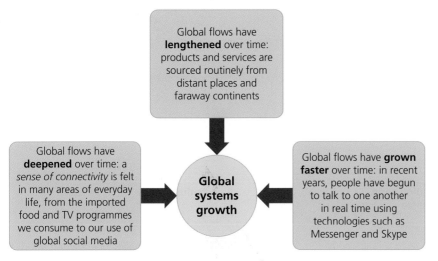

Global flows have **lengthened** over time: products and services are sourced routinely from distant places and faraway continents

Global flows have **deepened** over time: a *sense of connectivity* is felt in many areas of everyday life, from the imported food and TV programmes we consume to our use of global social media

**Global systems growth**

Global flows have **grown faster** over time: in recent years, people have begun to talk to one another in real time using technologies such as Messenger and Skype

◀ **Figure 1.1** Globalisation can be thought of as a series of linked changes *leading to the accelerated growth of global systems*: in recent decades, connections and flows between places, people and environments have significantly (i) lengthened, (ii) deepened and (iii) grown faster. This means inter-place interdependency has grown in scale, scope and intimacy

## Globalisation: a contested idea

As the section above indicated, the umbrella term 'globalisation' describes multiple processes of change which have resulted in places and people being more interconnected with one another than they used to be. Like the concept of 'development' (see Chapter 4), globalisation is such an expansive idea that it has become a widely contested term, meaning that people frequently disagree over its meaning and value. Important debates exist over (i) the historical accuracy of some accounts of globalisation, (ii) how globalisation is best defined and (iii) whether globalisation should be viewed as a positive or negative thing.

### Differing views about history and globalisation

When did globalisation begin? Opinions diverge about whether it is an entirely new phenomenon or not. Certainly, global system growth has been happening for millennia and modern globalisation has deep historical roots. It is the continuation of a far older, and ongoing, economic and political project of global trade and empire-building. Interdependence is not a new idea either. Since the time of the world's first great civilisations – such as ancient Egypt, Babylon and Rome – flows of people, goods and ideas have operated globally (see Figure 1.2). More recently, during the nineteenth century, the British Empire provided the UK's people and the English language with a global sphere of influence.

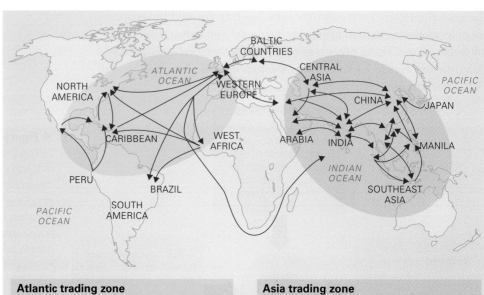

**Atlantic trading zone**

| | |
|---|---|
| Enslaved people | Furs |
| Wood | Gold and silver |
| Coffee | Sugar |
| Tobacco | Diamonds |

**Asia trading zone**

| | |
|---|---|
| Textiles | Porcelain |
| Shoes | Opium |
| Ceramics | Indian cotton |
| Tea | Mother-of-pearl |
| Clothing | Spices |
| Furniture | Silver |

▲ **Figure 1.2** Global trading zones in the early sixteenth to eighteenth centuries: how do these global systems differ from those of today?

🔑 **KEY TERMS**

**Colonialism** The establishment and maintenance of rule by a sovereign power over a subordinate country. By the end of the nineteenth century, the British Empire directly controlled one-quarter of the world and its people, for example.

**Deregulation** The process of reducing or eliminating government-imposed rules and restrictions on certain industries (such as financial services).

There is therefore nothing fundamentally new in global system growth resulting from the projection of global power and influence by strong individuals, nations and businesses. In the past, global connections were achieved through trade, colonialism and intergovernmental co-operation. For example, after the First World War ended in 1918 the League of Nations, a precursor to today's United Nations, was first established.

Modern globalisation does differ markedly in many respects from these earlier phases of world development, however. In particular, the sheer scale and number of flows of money, materials, ideas, information and people has changed, as Figure 1.2 shows. A sensible conclusion to arrive at is that the modern era of globalisation differs from what went before because of a *step change in connectivity* occurring in later decades of the twentieth century. That was when new communication technologies began to transform the way people lived in wealthier parts of the world. During the 1980s especially, the growth of the internet (see page 21) and deregulation of financial markets amplified and accelerated global flows significantly.

## Differing views about the meaning of globalisation

There is, it seems, little agreement about how globalisation should be defined. Many differing interpretations can be found and these vary markedly in their character, as Figure 1.3 shows. The words, tone and emphasis of the definitions used by different people and agencies reflect their own varying priorities.

- Some definitions are primarily economic, such as the statement used by the International Monetary Fund (IMF), while others acknowledge there are cultural dimensions to globalisation too.
- *Causality* can be identified in some of the statements, whereby globalisation has been presented as an *outcome* of technological or political changes.
- Some interpretations are deeply critical: this is because some people or organisations believe that recent global-scale changes have been very harmful to certain people, places and environments.

The definition that Malcolm Waters offered in his best-selling book *Globalization* (2000) is: 'A process in which the constraints of geography on economic, political, social and cultural arrangements recede, in which people are becoming increasingly aware that they [the constraints] are

| It's all about economics | It's a bad thing | It's caused by powerful forces |
|---|---|---|
| The term 'globalisation' refers to the increasing integration of economies around the world, particularly through the movement of goods, services, and capital across borders.<br><br>*IMF* | A rapid and huge increase in the amount of economic activity taking place across national boundaries. The current form of globalisation … has brought poverty and hardship to millions of workers, particularly those in developing countries.<br><br>*UK Trades Union Congress* | The world's economies have developed ever-closer links since 1950, in trade, investment and production. Known as globalisation, the changes have been driven by liberalisation of trade and finance, how companies work, and improvements to transport and communications.<br><br>*BBC* |
| It's a cultural thing | It benefits everyone | It's complex, so views differ |
| Globalisation might mean sitting in your living room in Estonia while communicating with a friend in Zimbabwe. It might be taking a Bollywood dance class in London. Or it might be symbolized by eating Ecuadorian bananas in the European Union.<br><br>*World Bank (schools website)* | The expansion of global linkages, organization of social life on a global scale, and growth of global consciousness, hence consolidation of world society.<br><br>*Frank Lechner, The Globalization Reader* | A process in which the constraints of geography on economic, political, social and cultural arrangements recede, in which people are becoming increasingly aware that they are receding, and in which people act accordingly.<br><br>*Malcolm Waters, Globalization* |

◀ **Figure 1.3** Different definitions of globalisation: each has a different 'character', reflecting the varying views, priorities or concerns of its author or audience

receding, and in which people act accordingly.' For geography students, this interpretation has much to recommend it. Several different dimensions of human activity are included but, just as importantly, Waters views globalisation as a process *that may give rise to any number of different local reactions to the changes it brings*. He is not seeking simply to brand globalisation overall as a 'good' or 'bad' thing. Instead, we are prompted to think critically about what actually happens 'on the ground' to people in varying local contexts.

## Differing views about the outcomes of globalisation

Views vary enormously on the desirability or otherwise of globalisation. This theme is returned to throughout this book, particularly in Chapter 6. One of the definitions in Figure 1.3 is highly critical: the Trades Union Congress (TUC) characterises globalisation as a process that 'has brought poverty and hardship' to millions of workers in developing countries. Globalisation is often blamed too for **deindustrialisation** and the impoverishment of **blue-collar workers** in inner cities of the developed world. Environmentalists will sometimes characterise globalisation first and foremost as a 'climate change culprit'.

In contrast, when Lechner mentions the 'growth of global consciousness, hence consolidation of world society' in Figure 1.3, his tone is approving. Globalisation supporters – sometimes called **hyperglobalisers** – cite evidence showing globalisation has lifted hundreds of millions of people out of poverty (see Chapter 4) while also bringing life-enhancing consumer goods, such as fridges, cookers and heaters, to billions more at affordable prices. In many countries, heavy industry has all but vanished and been replaced by less dirty and dangerous service sector employment. Progressive social and cultural norms – including equality for women and respect for LGBT (lesbian, gay, bisexual and transgender) communities – have diffused widely. Globalisation also fosters the concept of **global citizenship**, which may enhance humanity's prospects of ultimately tackling tough global challenges such as climate change.

In summary, the increased tension between supporters and detractors of globalisation is a defining issue of our times. New movements have gained in strength, and the successes of Donald Trump and the UK 'Leave' campaign (in the 2016 referendum on EU membership) are widely portrayed as 'anti-global' political movements whose time has come (see Figure 1.4). Chapters 5 and 6 explore these themes in greater depth.

▲ **Figure 1.4** New political movements suggest people in the USA and UK are becoming less convinced that globalisation is beneficial

# Analysing globalisation

To analyse something means to 'break it down' in a structured way. With a very broad concept such as globalisation, this can involve 'unpacking' its different dimensions. Two possible analytic frameworks might be adopted:

- Deconstructing globalisation into a series of component parts, each of which deals with a particular human impact or theme (economic globalisation, social globalisation and so forth).
- Thinking sequentially about the different types of global flow (people, merchandise, information and so on) that globalisation depends on.

Figure 1.5 combines both approaches to provide an analytical overview of globalisation.

The first approach to studying globalisation – 'unpacking' it into different component parts – involves sifting through many different causes and impacts of globalisation and categorising them. One commonly used method is to distinguish between economic, social, cultural and political globalisation.

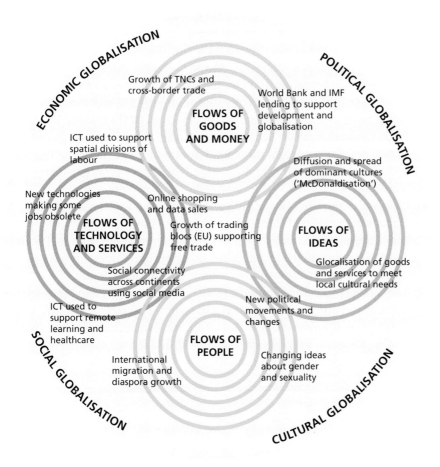

Pearson Edexcel AQA OCR WJEC/Eduqas

## KEY TERMS

**Deindustrialisation** A decrease in the importance of industrial activity in a local place or wider region, measured in terms of employment and/or output.

**Blue-collar workers** A term which is sometimes used to describe the workforce of manufacturing industries.

**Hyperglobalisation/ hyperglobalisers** The theory of hyperglobalisation proposes that the relevance and power of countries will reduce over time. Global flows of commodities and ideas may ultimately result in a shrinking and borderless world. Hyperglobalisers envisage a 'global village' where individual group attachments to ethnic and religious identity will be replaced by a shared identity based on the principles of global citizenship. There are, however, two competing perspectives on the desirability of this.

**Global citizenship** A way of living wherein a person identifies strongly with global-scale issues, values and culture rather than (or also alongside) narrower place-based identity.

◀ **Figure 1.5** The causes and effects of globalisation can be studied by (i) analysing economic, social, cultural and political aspects sequentially or (ii) analysing in turn how different global flows operate. This diagram combines both approaches

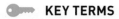

- *Economic globalisation*. This is the most important strand and the driving force behind global system growth. The expansion of large businesses in developing and **emerging economies** has 'enrolled' communities around the world into global systems as producers and/or consumers of their goods and services.
- *Cultural globalisation*. Widespread cultural changes continue to occur at a global scale alongside economic globalisation and the spread of consumerism. These include the diffusion and subsequent adoption of languages, fashions, music and foods that originated in powerful and influential states such as the USA, western European countries and Japan (see page 73). Cultural changes are sometimes fiercely resisted though (see page 74).
- *Social globalisation*. International migration has created extensive family networks straddling national borders. Global improvements in education and healthcare over time have resulted in higher literacy and life expectancy, although such changes are by no means uniform or universal. Social interconnectivity has grown over time too, thanks to the spread of social media.
- *Political globalisation*. Closely allied with the concept of **global governance**, the process of political globalisation includes the growth of large trading areas such as the European Union (EU) or the African Union. Important global organisations include the United Nations (UN), the 'G-groups' (the G7, G20 and G77) and the Organisation for Economic Co-operation and Development (OECD), all of which work to promote international or truly global growth and stability. The World Bank, the International Monetary Fund (IMF) and the World Trade Organization (WTO) have collectively helped create a globally shared political and legal framework for investment (see page 52), without which economic globalisation would have progressed far more slowly.
- *Links and connections*. In reality, the four different dimensions of globalisation outlined above are often intertwined like strands of rope. For example, world trade in manufactured commodities such as iPhones may lead in turn to the global diffusion of cultural values embedded in iPhone software: cultural and economic globalisation are thus inseparable processes. Similarly, there is a degree of overlap between political and economic forms of globalisation. This is demonstrated by the way a political ideology called **neoliberalism** influences the money-lending decisions of the World Bank and IMF (and so also affects patterns of economic globalisation).

## KEY TERMS

**Emerging economies**
Countries that have begun to experience higher rates of economic growth, often due to rapid factory expansion and industrialisation. Emerging economies correspond broadly with the World Bank's 'middle-income' group of countries and include China, India, Indonesia, Brazil, Mexico, Nigeria and South Africa.

**Global governance**
The term 'governance' suggests broader notions of steering or piloting rather than the direct form of control associated with 'government'. 'Global governance' therefore describes the steering rules, norms, codes and regulations used to regulate human activity at an international level. At this scale, regulation and laws can be tough to enforce, however.

**Neoliberalism** A management philosophy for economies and societies which takes the view that government interference should be kept to a minimum and that problems are best left for market forces of supply and demand to solve.

The second approach to studying globalisation – looking in turn at how different kinds of global flow operate – involves analysing ways in which different people and places are becoming increasingly interconnected as a result of flows of capital, people, merchandise, services and ideas.

## Capital (money) flows

Huge amounts of money cross national borders annually.

- Some of this – around US$2 trillion each year – is in the form of **foreign direct investment (FDI)** by large TNCs purchasing overseas assets (see page 27). Increasingly, **south–south** FDI flows have become an important part of global systems (see page 117).
- In 2017, the volume of foreign exchange transactions (which includes FDI) reached almost US$6 trillion, somewhat less than a peak of US$12.4 trillion which was reached in 2007–08 (see Figure 1.6).
- Vast flows of money are generated by investment banks, pension funds and private citizens who trade globally in shares and currencies for profit. This amounted to over US$70 trillion in 2016 according to one World Bank estimate.

**KEY TERMS**

**Foreign direct investment (FDI)** A financial investment made by a TNC or other international player (such as a government-controlled sovereign wealth fund) into a state's economy.

**South–south** In the context of global flows, this refers to movements of people, capital or trade from one part of the 'global south' (Asia, Africa, Latin America) to another. For example, Chinese investment in Sudan.

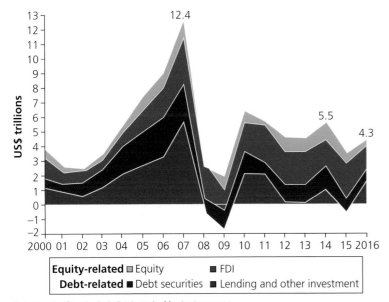

**Note:** Negative flows imply decline in stock of foreign investment

▲ **Figure 1.6** Cross-border capital flows 1990–2016: just before the global financial crisis (GFC) (see page 91), the figure peaked at over US$12 trillion in 2007

Other important capital flows include the following:

- *International lending and debt relief.* Loans can be an important financial flow for states at all levels of economic development. Sums borrowed by countries from the IMF and World Bank run into billions of dollars. Increasingly, China has become an important international lender too (see page 66).
- *International aid flows.* For example, flows of aid from the UK are directed towards Commonwealth countries. This is partly explained by the history the UK shares with its former colonies. Until very recently, India received more UK aid than any other country (on the grounds that half a billion Indians are still very poor and need help).
- *Remittances.* Around US$500 billion of remittances are currently sent home by migrants annually. This is three times the value of overseas development aid each year. Remittances can be transferred via banks or sent in the mail as cash. Unlike international aid and lending, remittances can be a peer-to-peer financial flow: money travels more or less directly from one family member to another. This cross-border money flow often plays a vital role in the social development of communities who have previously been excluded financially from access to education and healthcare.

### Flows of people (migrants and tourists)

Globalisation has led to a rise in migration flows both within countries (internal migration) and between them (international migration). A record number of people migrated internationally in 2015. There is now a total of more than 250 million people living in a country they were not born in. This represents between three to four per cent of the world's population. In fact, this percentage has not changed greatly over time despite the fact that the number of people migrating internationally has risen. This is because the total size of the world's population has grown too (between 1950 and 2019, world population grew from 4 billion to 7.6 billion).

Important changes have taken place in the *pattern* of international migration in recent years.

- As recently as the 1990s, international migration was directed mainly towards developed-world destinations such as New York and Paris. Since then, world cities in developing-world countries, such as Mumbai (India), Lagos (Nigeria), Dubai (UAE) and Riyadh (Saudi Arabia), have also begun to function as major global hubs for immigration, increasingly as a result of south–south movements (see page 127).
- Much international migration is relatively regionalised. In general, the largest labour flows connect neighbouring countries like the USA and Mexico, or Poland and Germany.

Global tourist flows continue also to rise year on year; some of the causes and consequences of this are explored in Chapter 5 (see pages 156–57).

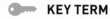 **KEY TERM**

**Global hub** A settlement (or wider region) providing a focal point for activities that have a global influence. All megacities are global hubs, along with some smaller cities such as Oxford and Cambridge, whose universities have a truly global reach. Global hubs are 'migration magnets'. In turn, migrant workers help power their economies.

## Merchandise flows (raw materials and manufactured goods)

In 2015, global gross domestic product (GDP) reached almost US$80 trillion in value. Of this, around one-third was generated by trade flows in agricultural and industrial commodities. In the past, raw material trade in food and minerals helped link states together. It remains important to global trade today, along with fossil fuel sales.

- Figure 1.7 uses proportional flow lines to illustrate how raw material trade flows expanded in size between 2000 and 2010. The grey lines (the inner lines drawn for each flow) show trade volumes in 2000, while the coloured lines show the increase by 2010.
- The reason for this heightened activity is the rapid development of emerging economies, especially China, India and Indonesia (combined, these countries are home to 3 billion people). Rising industrial demand for materials and increasing global middle-class consumer demand for food, gas and petrol are responsible for almost all growth in resource consumption across nearly every category shown.

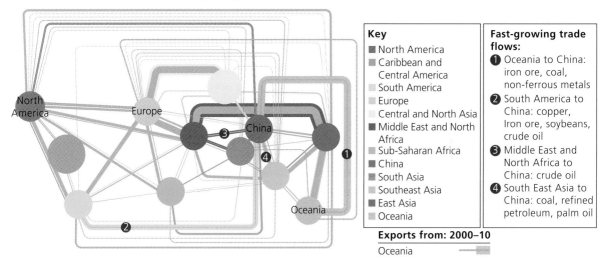

▲ **Figure 1.7** Growth in traded global flows of raw materials, 2000–2010
*Source: Chatham House*

Global flows of manufactured goods grew markedly in size during the 1990s and 2000s. The growth of trade in textiles and electronic goods was fuelled at first by low production costs in China and more recently by supply-chain growth in countries where rates of pay for workers are even lower, including Bangladesh, Vietnam and Ethiopia.

- Sixty years ago, the global trade pattern was very different. The majority of high-value manufactured goods were both produced and sold in North America, Europe, Japan and Australasia. Factories in these regions made use of raw materials imported from Asia, Africa and South America. Until more recently, these uneven trade flows contributed to the persistence of the so-called global 'north–south divide'.

- The subsequent rise of South Korea and later China (among others) as sites of innovation has transformed the global pattern of trade for manufactured goods. Korean electronics giant Samsung and China's Huawei have become major players in the production of telecommunications and home media devices. The geography of consumption for manufactured goods has altered beyond all recognition too. Over 1 billion mobile devices have been sold in India, for instance; it is also the world's fastest-growing car market. Large or fast-growing African economies, including Nigeria, South Africa, Egypt and Kenya, are increasingly viewed as important markets by Asian manufacturing companies, giving rise to greatly increased volumes of south–south trade.

## Flows of services (and the technologies that support them)

The rise of middle-class spending power in emerging economies has contributed to a rise in the volume of services which are traded. Regulatory and technological changes have also accelerated global growth in the following.

- *Tourism.* The value of the international tourist trade is widely believed to have doubled between 2005 and 2015 to a figure in excess of US$1 trillion (it is hard to make a precise estimate due to the many indirect benefits it creates). The number of international tourist arrivals has doubled in the same period and now exceeds 1 billion people. Much of the growth in activity has been generated by touristic movements within Asia. China now generates the highest volume of international tourism expenditure while Europe receives more tourist arrivals than any other continent.
- *Financial services and insurance.* Free-market liberalisation has played a major role in fostering international trade in financial services. For instance, the deregulation of the City of London in 1986 removed large amounts of 'red tape' and paved the way for London to become the world's leading global hub for financial services. Within the EU, cross-border trade in financial services has expanded in the absence of barriers. Large banks and insurance companies are able to sell services to customers in each of the EU's member states.
- *Online media and retailing.* One important recent development in global trade has been the arrival of on-demand media services. Faster broadband and powerful handheld computers have allowed companies like Amazon and Netflix to stream films and music on demand directly to consumers. Global delivery companies such as DHL have reaped the benefits of e-commerce growth.

Some countries and world cities serve as important global hubs for particular service flows. Important stock markets are found in New York, Tokyo, Shanghai and São Paulo. Nigeria and South Korea have successful television and film industries with many viewers in neighbouring countries.

### Global flows of ideas and information

The internet has brought real-time communication between distant places, allowing different ideas and perspectives to be shared. Facebook had 2 billion users in 2018; one effect of social media growth has been a worldwide sharing of views and concerns about issues ranging from climate change to animal welfare. Each user's own social network is a global system of sorts (see Figure 1.8). However, the same 'global' technology has also been used to destabilise democracies (page 30 examines how 'fake news' works) and, somewhat ironically, to foster local resistance against globalisation (see page 177).

## Analysing the forces which shape global flows

One view of global systems is that their growth is shaped by a combination of economic, technological and political forces (see Figure 1.9). The last of these – political forces – is the focus of Chapter 2, which deals with the way global organisations, agreements and powerful states have collectively shaped a worldwide framework of laws and norms which has (i) allowed global trade to flourish (and expand into new markets) and (ii) fostered the growth of global flows of people and ideas. The remainder of this chapter examines the constantly evolving technologies that have supported globalisation and the strategies TNCs have used to build their global businesses.

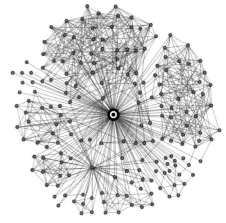

Total network: 179 friends

▲ **Figure 1.8** One person's Facebook friends visualised as a personal network. The linkages shown may cross national borders (the Facebook network of, say, an Indian economic migrant living in London is likely to comprise both friends and family in India along with newer acquaintances in England)

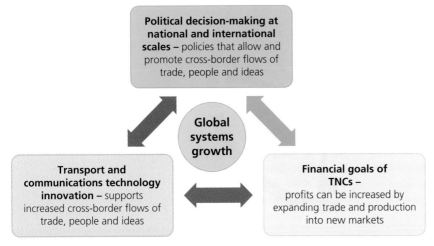

▲ **Figure 1.9** One view of the way global systems growth is shaped by the interaction of three different forces

# ANALYSIS AND INTERPRETATION

Figure 1.10 is a representation of global trade in 1992 and 2014. The circular flow diagrams show trade flows within (intra-regional) and between (inter-regional) different world regions. The colour shows which region each trade flow came from; the numbers give the total value of trade between and within regions.

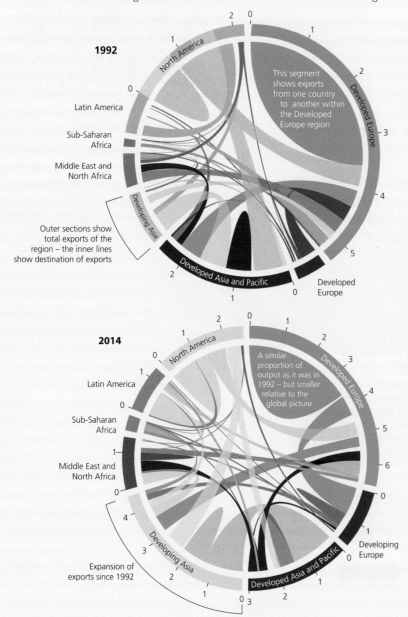

▲ **Figure 1.10** Changing inter-regional and intra-regional patterns of global trade, 1992–2014: the value of exports of goods are shown by region as a percentage of total world GDP. *Adapted from FT Graphic Alan Smith, Keith Fray; Sources: IMF DOTS; FT research*

(a) Estimate the percentage of global GDP in 1992 which is derived from intra-regional trade within the 'Developed Europe' region.

## GUIDANCE

Each world region is shown to be connected with other world regions via trade flows whose value can be estimated (as a percentage of global GDP). In addition, there are unconnected segments. These show the value of trade taking place among countries *within* each world region. The large blue segment shown on the diagram (1992 values) is our focus here. Can you estimate its value, using the scale?

(b) Analyse Figure 1.10 for evidence of globalisation.

## GUIDANCE

Every picture tells a story, and this pair of diagrams is no exception. Most definitions of globalisation use phrasing such as 'the increasing integration of economies' or 'accelerating connectivity'. In other words, globalisation has temporal and spatial dimensions – it is widely viewed as something which is 'speeding up' while at the same time also expanding *spatially* in its influence. And that is exactly what these diagrams show. If you compare 2014 with 1992, there are a greater number of different strands and connections: more and more places are becoming connected with one another, resulting in a more densely networked world. In other words, this illustration is a visualisation of what globalisation looks like.

(c) Assess the value of Figure 1.10 as a way of illustrating changes in global connectivity over time.

## GUIDANCE

An assessment of the value of a graph or chart can be carried out in two stages. First, we must ask: does it actually show what it claims to be showing? Second, has the information been presented in a way that helps an audience understand and engage with the issues? Following from this, two important criticisms of Figure 1.10 are that:

1 it only shows flows of trade (thus, it is only a partial representation of a far more complex process that has social, cultural and political processes)
2 circular flow diagrams are not particularly easy to understand and require careful explanation, which perhaps limits their usefulness (moreover, the data are shown as percentages rather than actual numbers, meaning we have no way of knowing whether the *actual volume* of global trade has risen or fallen over time).

A balanced assessment of source material should also acknowledge the strengths of the presentation methods used and not merely criticise them. One strength of Figure 1.10 lies in the way increasing global interconnectivity over this time period is very clearly communicated to an audience.

# CONTEMPORARY CASE STUDY: THE KOF GLOBALISATION INDEX

Some institutions and organisations have attempted to measure globalisation, but doing so is not straightforward due to its diverse character, and conflicts about what appropriate measures exist. Since 2006, the KOF Swiss Economic Institute (in Zürich, Switzerland) has attempted to quantify each country's level of globalisation on an annual basis. This analysis uses a multi-strand model of globalisation which recognises its 'different dimensions and characteristics'. KOF defines globalisation as:

*The process of creating networks of connections among actors at intra- or multi-continental distances, mediated through a variety of flows including people, information and ideas, capital, and goods. Globalisation is a process that erodes national boundaries, integrates national economies, cultures, technologies and governance, and produces complex relations of mutual interdependence.*

How then best to measure this? A complex methodology informs each report (see Table 1.1). Levels of economic globalisation are calculated by examining trade, FDI figures and any restrictions on international trade. Political globalisation is also factored in, for instance by counting how many embassies are found in a country and the number of UN peace missions it has participated in. Finally, social globalisation is accounted for, defined by KOF as 'the spread of ideas, information, images and people'. Data sources for this include levels of internet use, TV ownership and imports and exports of books. In 2018, Belgium and the Netherlands were the world's two most globalised countries according to the KOF Index, with both scoring 90 out of a theoretical maximum of 100 (see Figure 1.11).

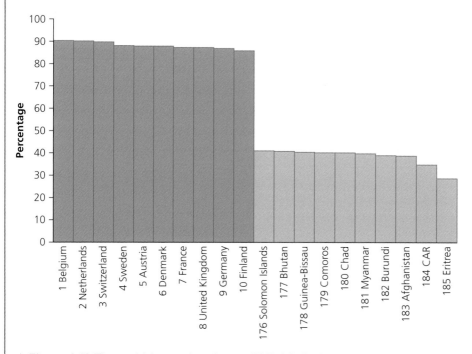

▲ **Figure 1.11** The ten highest and ten lowest KOF globalisation rankings for 2018

| Stage | |
|---|---|
| 1 | To study economic globalisation, information is collected showing long-distance flows of goods, capital and services. Data on trade, FDI and portfolio investment are studied, including figures from the World Bank. |
| 2 | A second measure of economic globalisation is made by examining restrictions on trade and capital movement. Hidden import barriers, mean tariff rates and taxes on international trade are recorded. As the mean tariff rate increases, countries are assigned lower ratings. |
| 3 | Political globalisation data are looked at next. To proxy the degree of political globalisation, KOF records the number of embassies and high commissions in a country as well as the number of international organisations the country is a member of and also the number of UN peace missions a country has participated in. |
| 4 | The assessment of social globalisation starts with use of personal contacts data. KOF measures direct interaction among people living in different countries by recording (i) international telecom traffic (traffic in minutes per person), (ii) tourist numbers (the size of incoming and outgoing flows) and (iii) the number of international letters sent and received. |
| 5 | The study of social globalisation also requires use of information flows data. World Bank statistics are employed to measure the potential flow of ideas and images. National numbers of internet users (per 1000 people) and the share of households with a television are thought to show 'people's potential for receiving news from other countries – they thus contribute to the global spread of ideas'. |
| 6 | Finally, 'cultural proximity' data are collected. By KOF's own admission, this is the dimension of social globalisation that is most difficult to measure. The preferred data source is imported and exported books. 'Traded books', explains KOF, 'proxy the extent to which beliefs and values move across national borders.' |
| 7 | Now all the data have been collected, a total of 24 variables (covering economic, political and social globalisation) are each converted into an index on a scale of 1 to 100, where 100 is the maximum value for each specific variable over the period 1970–2006 and 1 is the minimum value. High values denote greater globalisation. However, not all data are available for all countries and all years. Missing values are substituted by the latest data available. Averaging then produces a final score out of 100. |
| 8 | Each year's new KOF Globalisation Index scores are added to a historical series covering more than 30 years, beginning in 1970. Changes in globalisation over time can then be studied. |

▲ **Table 1.1** Calculating's a country's globalisation score using the KOF Globalisation Index

While there is obvious merit in KOF's multi-strand approach to measuring globalisation, there are considerable grounds for criticism too. For instance, does possession of a television really make a family more globalised? Equally, the reasons why some countries volunteer large numbers of troops for UN missions are complex; it need not necessarily be is not the case that the most economically globalised countries are *always* the most militarily pro-active (take Japan and Germany, for instance).

## KEY TERMS

**Containerisation** The practice of transporting merchandise in large containers. Intermodal containers are large-capacity storage units that can be transported long distances using multiple types of transport, such as shipping and rail, without the freight being taken out of the container.

**Artificial intelligence (AI)** Forms of intelligence and learning shown by computers, ranging from speech recognition to complex problem-solving. Many kinds of employment are believed to be threatened by near-future advances in AI capabilities.

**Shrinking-world effect** Heightened connectivity changes our conception of time, distance and potential barriers to the migration of people, goods, money and information. Distant places feel closer than in the past. As a result, the definition of what constitutes a 'near' or 'far' place changes in line with shifting perceptions of spatial relations. The shrinking world is a very important idea in A-level and undergraduate geography because it forms a 'synoptic bridge' between the study of global systems and changing places.

# ② Global linkages created by technology

▶ *What role have transport and communications played in the growth of global systems and a shrinking world?*

## Transport, technology and time–space compression

The structural changes shown by Figure 1.1 – *longer, deeper and faster connections helping to link together different places, people and environments* – were enabled in part by technological changes over time. Developments in transport and trade during the nineteenth century (notably railways, the telegraph and steamships) accelerated in the twentieth century with the arrival of jet aircraft and **containerisation**. The twenty-first century's ever-more complex and deeply interconnected communications systems depend on highly sophisticated technologies supported by global networks of fibre-optic cables. We also live in a world where management, information and security systems increasingly rely on **artificial intelligence (AI).**

Collectively, these technologies contribute to an ongoing process called *time–space compression* (see page 1) which leads individuals and societies to experience the above-mentioned **shrinking-world effect**. The geographer Doreen Massey devoted much of her career to understanding how people's sense of place

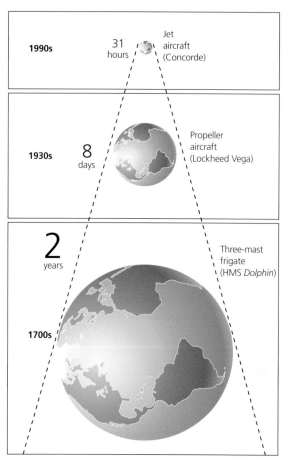

▲ **Figure 1.12** A shrinking world: the changing time taken to circumnavigate the world

and identity has changed on account of the way places have become more interconnected over time. For people living in the UK in the twenty-first century, the perception of what feels like a 'near' or 'far' place is very different from the way the world was experienced by most Victorians, for example. Even in the mid-1800s, ordinary people knew little of the world beyond their immediate neighbourhood or city. This was because travelling long distances was still a prohibitively expensive and slow process. Since then, successive rounds of technological and transport innovation have chipped away minutes, hours and days from the time it takes to travel to other places or communicate with people in distant countries (see Figure 1.12). The resulting experience is called time–space compression (see page 1).

## Thinking critically about technology

We must not get *too* carried away by the idea that 'everybody' now experiences the same sense of a shrinking world, however. In her own books and journal articles, Massey was critical of writing that implies international travel in aeroplanes has become a routine experience for humanity. Billions of people remain *weakly* connected to global systems. Some societies are barely connected at all, such as the subsistence farmers of Kenya's Mau Forest (see page 158). Although it is true that the majority of people in the continent of Africa had access to mobile phones in 2018, most still did not use them to access the internet (see page 23).

It is also important to reflect on whether the relationship between global systems and technology is perhaps more complex than many narratives suggest. Transport and communications are widely understood to be 'drivers' of globalisation. But this is a reductive view of causality, however, if it ignores another side to the story: that new technologies are also an *outcome* of globalisation. This important idea – that technological change is simultaneously *both a cause and an effect* of globalisation – is the focus of this chapter's final debate (see page 31).

## Transport developments over time

Table 1.2 shows four especially significant post-war transport innovations that have helped increase spatial interactions between places.

**KEY TERM**

Subsistence Economic self-sufficiency, often practised in traditional agricultural communities living close to the poverty line. Only enough food and materials are produced to meet the immediate needs of a family or small community. There is negligible surplus for trade and so a market economy does not develop.

| Container shipping | According to one estimate, around 200 million individual container movements take place each year. Although this number may have fallen in recent years, shipping has remained the 'backbone' of the global economy since the 1950s. Everything from tea bags to trampolines can be transported efficiently across the planet using intermodal containers. The South Korean-built vessel *OOCL Hong Kong* is 400 metres long, nearly 60 metres wide and can carry over 21,000 containers. |
|---|---|

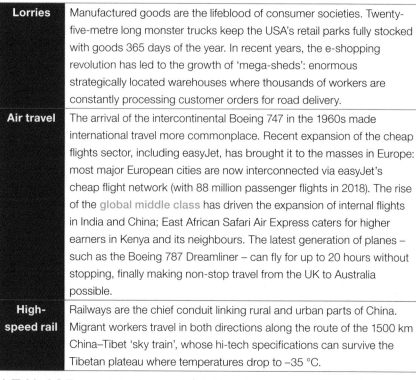

**KEY TERM**

**Global middle class**
Globally, the middle class is defined as people with discretionary income they can spend on consumer goods. Definitions vary: some organisations define the global middle class as people with an annual income of over US$10,000; others use a benchmark of US$10 per day income.

| | |
|---|---|
| **Lorries** | Manufactured goods are the lifeblood of consumer societies. Twenty-five-metre long monster trucks keep the USA's retail parks fully stocked with goods 365 days of the year. In recent years, the e-shopping revolution has led to the growth of 'mega-sheds': enormous strategically located warehouses where thousands of workers are constantly processing customer orders for road delivery. |
| **Air travel** | The arrival of the intercontinental Boeing 747 in the 1960s made international travel more commonplace. Recent expansion of the cheap flights sector, including easyJet, has brought it to the masses in Europe: most major European cities are now interconnected via easyJet's cheap flight network (with 88 million passenger flights in 2018). The rise of the global middle class has driven the expansion of internal flights in India and China; East African Safari Air Express caters for higher earners in Kenya and its neighbours. The latest generation of planes – such as the Boeing 787 Dreamliner – can fly for up to 20 hours without stopping, finally making non-stop travel from the UK to Australia possible. |
| **High-speed rail** | Railways are the chief conduit linking rural and urban parts of China. Migrant workers travel in both directions along the route of the 1500 km China–Tibet 'sky train', whose hi-tech specifications can survive the Tibetan plateau where temperatures drop to –35 °C. |

▲ **Table 1.2** Transport development has led to a shrinking world. But which technologies have had the greatest impact?

## Global data flow patterns and trends

Alongside transport improvements, information and communications technology (ICT) has transformed the way people interact, work and consume services and entertainment. Important milestones in data transfer, storage and retrieval technologies are shown in Table 1.3. The forces behind this technological revolution are varied.

- Some important developments have been driven by government funding of the military. The internet began life as part of a scheme funded by the US Defence Department during the Cold War. Early computer network ARPANET was designed during the 1960s as a way of linking important research computers in just a handful of different locations. Since then, connectivity between people and places has grown exponentially.
- Other crucial breakthroughs came from electronics hobbyists and university researchers: the modem device which links two computers together via conventional telephone lines (without going through a host system) was developed by two Chicago students. This vital stepping stone in the evolution of the internet came about as a result of their determination to avoid going outside during the cold Chicago winter of 1978.

- Increasingly, innovation is driven by the need of TNCs to protect their market share. Samsung, Apple, Huawei and other electronics companies constantly refine their products in a competitive marketplace that quickly becomes 'saturated'. This means sales fall after most people have purchased the latest device. Therefore, in order to maintain profits, companies must regularly create a newer, improved product to sell to customers, ideally in a very short timeframe of just one or two years.

| Telephone and the telegraph | ■ The first telegraph cables across the Atlantic in the 1860s finally replaced a three-week boat journey with instantaneous communication. For the first time, it became possible for people living in one part of the world to know what was happening in other places at that same moment. This was a truly revolutionary moment in human history.<br><br>■ The telephone, telegraph's successor, remains a core technology for communicating across distance. |
|---|---|
| Personal computers and 'the internet of things' | ■ The microprocessor was introduced by Silicon Valley's Intel corporation in 1971; soon afterwards, small-scale computers began to be designed around microprocessors, including early Apple microcomputers designed by high-school drop-outs Steve Wozniak and Steve Jobs in Silicon Valley.<br><br>■ User-friendly interface technology and software was introduced first to the Apple Macintosh in 1984 and to the PC by Microsoft as Windows 1.0 in 1985.<br><br>■ Computers have evolved into laptops, tablets and small handheld devices. Small networked computers are increasingly integrated seamlessly into cars and even fridges. This new age of smart devices is sometimes called 'the internet of things'. |
| Broadband and fibre optics | ■ With the roll-out of broadband internet in the 1980s and 1990s, large amounts of data could be moved quickly through cyberspace. Today, enormous data flows are carried by ocean-floor fibre-optic cables owned by national governments or TNCs such as Google. More than 1 million kilometres of flexible under-sea cables with a similar diameter to that of garden hoses carry all the world's emails, searches and tweets. The pattern of data flow is shown in Figure 1.13.<br><br>■ Telecoms network-builders have had to overcome some enormous physical challenges. They have slung mile upon mile of vulnerable fibre-optic cable across the vast plains of the ocean floor. This in turn has created new economic risks for societies: in 2006, a major submarine earthquake and landslide destroyed Taiwan's telecom link with the Philippines, disrupting TNC operations. Cyclones and tsunamis destroy cables; so too can dropped anchors. |
| GIS and GPS | ■ The first global positioning system (GPS) satellite was launched in the 1970s. There are now 24 situated 10,000 km above the Earth. These satellites continuously broadcast position and time data to users throughout the world.<br><br>■ Geographic Information Systems (GIS) are a collection of software systems that can collect, manage and analyse satellite data. |

▲ **Table 1.3** Important elements of the growth over time of the communications technologies on which global system growth depends

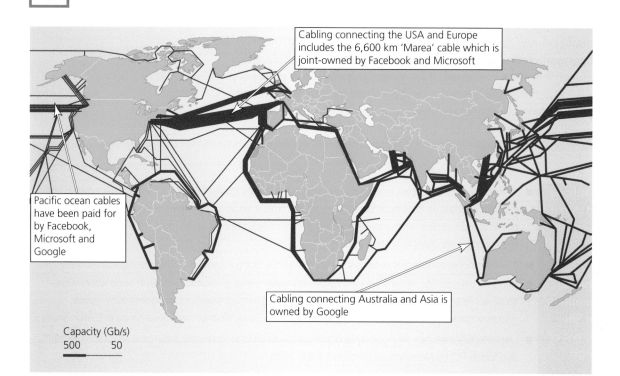

Cabling connecting the USA and Europe includes the 6,600 km 'Marea' cable which is joint-owned by Facebook and Microsoft

Pacific ocean cables have been paid for by Facebook, Microsoft and Google

Cabling connecting Australia and Asia is owned by Google

Capacity (Gb/s)
500    50

▲ **Figure 1.13** The global pattern of data flows carried by under-sea fibre-optic cables. In the past, submarine communications infrastructure was funded by governments; increasingly, large TNCs such as the FANGs (Facebook, Amazon, Netflix and Google) now foot the bill

## The mobile phone revolution and electronic banking in developing countries

A lack of communications infrastructure used to be a big obstacle to economic growth for developing countries. Now, however, mobile phones are changing lives for the better by connecting people and places. The scale and pace of change is extraordinary.

- In 2005, six per cent of all Africans owned a mobile phone. By 2015 this had risen tenfold to 60 per cent due to falling prices and the growth of provider companies such as Kenya's Safaricom. In 2018, only ten per cent of Africa's population lived in areas where no mobile service is available.
- Rising uptake in Asia (in India, over 1 billion people are mobile subscribers) means since 2016 there have been more mobile phones than people on the planet.

In 2007, Kenya's Safaricom introduced M-Pesa, a simple mobile phone service that allows credit to be directly transferred between phone users. Within a decade of this launch, over 20 million Kenyans were using the services to access their bank accounts or send peer-to-peer money payments to one another's mobile phones. This has revolutionised life for local individuals and businesses.

- The equivalent of more than one-half of Kenya's GDP is now sent through the M-Pesa system annually.
- People in towns and cities use mobiles to make payments for utility bills and school fees.
- In rural areas, fishermen and farmers use mobiles to check market prices before selling produce.
- Women in rural areas are able to secure **microloans** from development banks by using their M-Pesa bills as proof that they have a good credit record. This new ability to borrow is playing a vital role in lifting rural families out of poverty (see page 113).

Recently, M-Pesa has spread more widely throughout East Africa to become an international phenomenon. In most countries, the proportion of mobile users with internet access (rather than simply text messaging) is growing (see Figure 1.14). As a result, more people throughout the world are gaining access to apps for healthcare, education, finance, agriculture, retail and government services. Already, there is an East African app for almost everything: herding cattle in Kenya (i-Cow), private security in Ghana (hei julor!) and remotely monitoring patients in Zimbabwe (Econet). In Uganda, a mobile service called Yoza connects people with dirty laundry to mobile washerwomen.

 **KEY TERM**

**Microloans** Affordable credit that can, for example, help farmers grow a bigger crop and sell the surplus for cash.

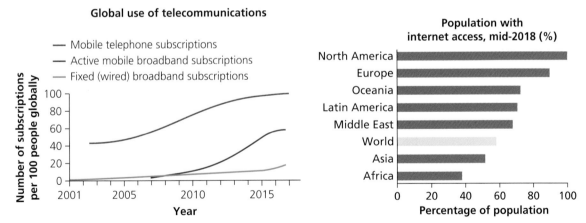

▲ **Figure 1.14** Global trends and patterns for communications technology use, 2018

# 3 Global linkages created by trade and transnational corporations

▶ *What role have trade and transnational corporations played in globalisation and global systems growth?*

## Global trade and global flows

International trade is the cornerstone of global systems and globalisation is inextricably bound up with the expansion of capitalism on a planetary scale. Transnational corporations (TNCs) are the world's most important global system-builders and this section is devoted to studying how they connect people, places and environments together. But, just like house-builders in a local context, TNCs can only operate *when planning laws allow it*. Businesses are not all-powerful entities and their hands are often tied by rules and regulations. A TNC's freedom to invest in new markets can be dependent on political systems at both national and international levels. You will find out more about this in Chapter 2's analysis of the multi-scalar legal frameworks within which TNCs must operate when building their global businesses.

TNCs connect places at a global scale in direct and indirect ways via flows of money, merchandise and services.

● Company managers may take the decision to invest money in new production sites or to market goods and services for new emerging markets. These flows are planned strategically and controlled directly by business operations managers.

● Additionally, TNCs connect places together through flows of ideas and information, which can bring technological and cultural change on a global scale. These flows differ from investment and trading because they are, for the most part, unintentional and indirect side effects of economic globalisation. For example, US films and TV shows are streamed around the world by TNCs like Netflix and YouTube. At the same time as flows of money are being generated by the consumers of these services, flows of ideas are occurring too. For example, US-produced media services typically offer a relatively liberal and progressive view of the world. Many shows feature women and LGBT actors in leading roles. TV shows may therefore help change social attitudes in countries where greater progress towards equality is still needed. This theme is returned to in Chapter 4.

# TNCs and their spatial networks and strategies

The largest TNCs create networks of connected countries. Company investments, trade flows and organisation structures all help to link different places together. Complex forms of interconnectivity sometimes develop, particularly for companies operating extensive supply chains or sourcing goods from thousands of different places.

Global patterns of production, distribution and consumption depend on enormous capital flows in the form of foreign direct investment (FDI).

- The great global increase in interconnectivity and interdependency seen over the last 30 years owes more to FDI than any other type of monetary flow, with offshore investment by TNCs rising near exponentially from US$20 billion in 1980 to a peak of around US$2 trillion in 2007–08 (before falling to US$1.2 trillion by 2015–16 because of the fragility of the world economy during the 2010s).
- There are perhaps as many as 100,000 TNCs (when defined as companies with 'operations in more than one country'). The top 100 of these own 20 per cent of world financial assets, employ 6 million people and enjoy 30 per cent of global consumer sales.

The worldwide distribution of TNC headquarters has changed in recent years as more companies from emerging economies have become major investors in global markets (see Figure 1.15). Enormous outbound foreign investments are made each year by India's CFS and Infosys, as well as China's Haier and Huawei, for example. The 2006 buy-out of Corus, previously British Steel, by Indian firm Tata was a landmark move some commentators branded 'reverse colonialism'. There are many Indian TNC owners – such as Tata chief Lakshmi Mittal – in the 'top 100 world billionaire rich list'.

In which nation is the bulk of assets and senior staff located?

In which nation is the TNC as a whole taxed on its worldwide earnings?

What is the nationality of the board of directors and decision-makers?

To which nation would the group turn for diplomatic protection and support?

What is the legal nationality of the parent company?

▲ **Figure 1.15** Investigating the global distribution of TNCs: these are questions you can ask when investigating the 'national identity' of TNCs

**KEY TERM**

**Glocalisation** This term is used sometimes to describe the local sourcing of parts by TNCs when establishing themselves overseas in places where they make or assemble merchandise close to markets. At the same time, they are able to customise their products to meet local tastes or laws in an attempt to further boost sales. Media and publishing TNCs often glocalise the content of TV shows, magazines, websites and other information services.

# Foreign direct investment strategies used by TNCs

A range of investment strategies are used by TNCs to build their global businesses, with the precise course chosen very much depending on the industrial sector the company operates within (see Table 1.4). Mining and oil companies and food-producing agribusinesses do not deal in products or services that can be, or need to be, customised for different markets. They expand into new territories primarily by buying up other companies instead. In contrast, manufacturing and banking companies, such as Unilever, Samsung and Citigroup, will often modify their products and services for the diverse markets they seek to conquer, using a strategy called glocalisation (see Figure 1.16).

## Outsourcing and global production networks

As an alternative to investing in its own overseas operations, a TNC can choose instead to offer a working contract to another foreign company: this is called *outsourcing*. In recent decades, China and India have become major outsourcing destinations for manufacturing and services respectively (see Chapter 4, pages 117 and 129).

▲ **Figure 1.16** Glocalisation: Nestlé produces numerous flavoured varieties of KitKat® customised especially for the Japanese market

| Strategy | Definition and example | Evaluation |
|---|---|---|
| Offshoring | ■ This involves TNCs moving parts of their own production process (factories or offices) to other countries to reduce labour or other costs. For instance, the UK technology company Dyson moved its own manufacturing division to Malaysia in 2002; further bases have since been added in China, the Philippines and Singapore.<br><br>■ For large media and finance companies, it can involve setting up new international offices. UK accounting and consultancy firm KPMG now has offices in over 100 states.<br><br>■ By offshoring, TNCs can locate factories and offices closer to the markets they will be serving. Japanese car companies like Nissan have built factories inside the EU in order to serve European markets directly (and also avoid EU import taxes). | ■ There are many benefits of creating a spatial division of labour, both for the TNC (whose profits rise) and the states it invests in.<br><br>■ However, offshoring may have costs too. Job losses in the TNC's country of origin may result from the relocation of manufacturing overseas. This can attract criticism, leading to poor sales.<br><br>■ Companies may expose themselves to a range of new political and physical risks by investing in certain states. In recent years there has been a growing trend of TNCs reshoring some operations in order to reduce these risks (see Chapter 6).<br><br>■ Firms must also think carefully about how much decision-making power to grant to their different overseas operations. |

| Strategy | Definition and example | Evaluation |
|---|---|---|
| Acquisitions | ■ When an international corporate merger takes place, two firms in different countries join forces to create a single entity. The energy TNC Royal Dutch Shell has a 'dual-listed' corporate structure, meaning it maintains headquarters and pays corporation tax in both the UK and the Netherlands.<br><br>■ When a TNC launches a **takeover** of a company in another country, it is called a foreign acquisition. In 2010, UK chocolate-maker Cadbury was taken over by US food giant Kraft, for instance. Revenue from Cadbury's UK sales now feeds the profits of US-registered firm Mondelēz International (a newly formed division of Kraft). | ■ In 2015, around one-third of all FDI consisted of cross-border mergers and acquisitions. Clearly, these are very important strategies which TNCs benefit from in numerous ways, including expanded markets and the opportunity to reduce costs (and therefore increase profits) through rationalisation (streamlining operations).<br><br>■ Changes in TNC ownership affect the geography of global financial flows. Large profit flows are redirected towards the state where the buyer is headquartered. This has major implications in turn for states and societies because of the financial losses or gains in corporation tax paid to governments. |
| Joint venture (JV) | ■ This involves two companies forming a partnership to handle business in a particular territory (but without actually merging to become a single entity).<br><br>■ TNCs must sometimes set up joint ventures because local investment law demands it, for instance in India.<br><br>■ In North India, McDonald's restaurants are part-owned by Vikram Bakshi's Connaught Plaza Restaurants. The local success of this venture owes much to glocalisation strategies (see below) developed by the partnership, such as the introduction of the vegetarian McAloo Tikki Burger. It makes good business sense for the TNC to work with a local company that has a better understanding of local customer preferences (see also page 181). | ■ Setting up a joint venture reduces the risk a TNC is exposed to; it must also share the profits though.<br><br>■ Global financial flow patterns are affected by JVs. In the case of McDonald's, only a share of the profits is transferred to the USA; the rest stays in India.<br><br>■ The combined expertise of a global TNC working with a local company can make the venture more successful than either stakeholder would be able to achieve working alone. The logistics of attempting to do business in many different countries creates headaches for large corporations. The partner company's local knowledge is therefore valuable when trying to gain a toehold in a lucrative new emerging market (India's retail market was worth US$1 trillion in 2018). |
| Glocalisation | ■ TNCs sometimes invest in new product designs as part of their overseas investment strategy. Glocalisation involves adapting a 'global' product to take account of geographical variations in people's taste, religion and interests.<br><br>■ This strategy is examined in greater depth in Chapter 2. | ■ It makes business sense for some TNCs to pay attention to their customers' culture. However, not all companies need to glocalise products. For some big-name TNCs like Lego, the 'authentic' uniformity of their global brand is what generates sales. For oil companies, glocalisation has little or no relevance for their industrial sector. |

▲ **Table 1.4** Different TNC investment strategies that help build global systems

**Spatial division of labour**
The common practice among
TNCs of moving low-skilled
work abroad (or 'offshore')
to places where labour
costs are low. Important
skilled management jobs
are retained at the TNC's
headquarters in its country
of origin.

**Reshoring** Also known as
onshoring and backshoring,
this involves a TNC
abandoning lengthy supply
chains and instead returning
productive operations
to the country where it
is headquartered. The
company will no longer make
use of a spatial division of
labour.

**Takeover** When one TNC
seizes control of another,
having bought shares or
persuaded shareholders to
accept the acquisition offer.

**Global production
network (GPN)** A chain of
connected suppliers of parts
and materials that contribute
to the manufacturing or
assembly of consumer
goods. The network serves
the needs of a TNC, such as
Apple or Tesco.

**Indigenous people** Ethnic
groups who have enjoyed
the uninterrupted occupation
of a place for long periods of
time (pre-dating any arrival of
more recent migrants).

Outsourcing frees the TNC from the hassle of building or leasing property
and people directly. However, it also introduces new elements of risk into
the supply chain (see also Chapter 6, page 193). A TNC may struggle to
closely monitor the production of goods or services by a supplier. This could
jeopardise both the quality of the products and the working conditions of
the people who make them. Both of these issues can affect brand
reputation. The 2013 collapse of the unsafe Rana Plaza building in Dhaka,
Bangladesh, led to the deaths of 1,100 textile workers. It was also deeply
troubling for Walmart, Matalan and many other major TNCs who regularly
outsourced clothing orders to Rana Plaza (see page 151).

The resulting geography of many large TNCs consists of a complex web of
combined offshored and outsourced operations which, in turn, serve many
different worldwide markets. The resulting series of arrangements is called
a global production network (GPN). A TNC manages its GPN in the
same way the captain of a team manages other players. As globalisation has
accelerated, so too have the size and density of global production networks
spanning food, manufacturing, retailing, technology and financial services.
Food giant Kraft and electronics firm IBM both have 30,000 suppliers
providing the ingredients they need. GPNs are looked at in more depth –
along with the interdependent relationships between different players and
places they give rise to – in Chapter 3 (see page 78).

## The uneven geography of TNCs and global production networks

TNCs link together different places and environments through their supply
chains and marketing strategies. But some parts of the world have received
far more FDI from TNCs than other areas.

- Not all places have sufficient market potential to attract large retailers
  yet. For reasons of poverty, paucity of population (perhaps because of
  physical isolation) or cultural differences, some places remain relatively
  'switched-off' to the global flows generated by TNCs. One example of
  this is Papua New Guinea's tropical rainforest societies, who are among
  the world's last isolated groups of indigenous people (see Chapter 5).
- In contrast, the strengthening of Latin American, Asian and Middle
  Eastern economies has prompted an explosion of TNC interest in these
  emerging markets, where over 2 billion people have moved from poverty
  into higher-income brackets since 1990.

# CONTEMPORARY CASE STUDY: FACEBOOK

Facebook is a leading media TNC and an important architect of social and cultural globalisation. In recent years, its social network has ballooned in size and influence: in 2018, there were over 2.2 billion Facebook users (see Figure 1.17). Facebook services allow users to build a global network of personal contacts, making the company a major contributor to the shrinking-world effect.

- Facebook has grown its global market share in part through the acquisition of more than 50 other companies since 2005, including Instagram and WhatsApp.

- The company's explosive growth has been aided by innovation from other technology companies whose ever-improving smartphone designs have brought social networks to hundreds of millions of new consumers in just a few years.

To manage its global operations, Facebook has established over 70 regional offices: in 2018, there were 24 in North America, 18 in Europe, the Middle East and Africa; 16 in Asia-Pacific and 4 in Latin America. Facebook FDI into other states takes two forms.

1 *Regional offices where sales strategies and advertising campaigns are developed.* There are major offices in London and Singapore, for instance. Its Indian operations are especially important: India is the largest market for Facebook outside the USA, with close to 220 million monthly active users in 2018.

2 *Regional data-storage facilities called server farms.* These are storage facilities that are filled with cupboard-sized racks of computer servers (giant hard drives) that store and move data such as photos, films and music. Facebook's US$760 million major data centre in Luleå is located in Sweden's coldest region. The low temperatures reduce the cost of cooling the tens of thousands of hard drives installed at the 30,000 square metre facility (the size of 11 football pitches). A flat, glaciated valley floor at Luleå provides plenty of room for future expansion,

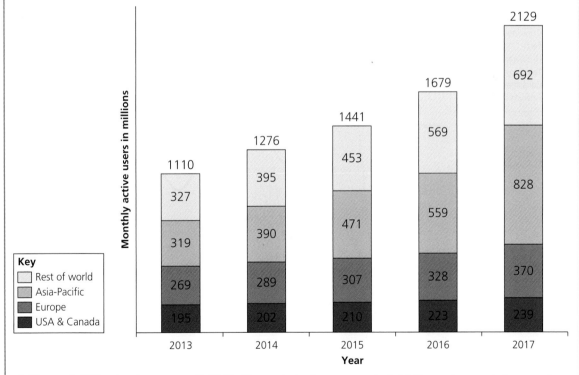

▲ **Figure 1.17** Facebook user growth 2013–17: over just a four-year period, 1 billion extra people began using this company's services

making this an interesting case of physical geography influencing FDI flows. Luleå also provides Facebook with a source of renewable energy: hydroelectric power. The TNC has been working with Greenpeace to develop a more sustainable business model.

Facebook's relentless global expansion has recently been called into question. Evidence suggests younger Millennials and school children are abandoning the platform in favour of other social media sites. In 2018, more than US$120 billion was wiped off the value of

Facebook shares in a single day, marking the biggest one-day 'value destruction' of a listed company in US history, almost equal to the entire combined value of McDonald's and Nike. Around the world, national governments are becoming more critical of (i) how some people misuse Facebook to exploit vulnerable people (such as grooming and radicalisation) and (ii) the alleged interference in US and European elections (involving carefully-targeted Facebook messages sent from Russian 'fake news farms'). Facebook began to diversify further in 2017 with the release of Facebook Watch, an online streaming service.

## ANALYSIS AND INTERPRETATION

Study Figure 1.18, which shows the changing rank order of the USA's largest TNCs between 2009 and 2018.

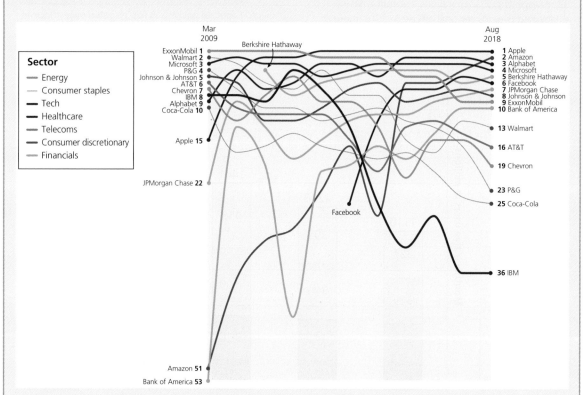

▲ **Figure 1.18** Changes in the rank ordering of US TNCs according to their total market value as businesses. The companies are colourcoded according to which industrial sector they are most closely associated with.
*FT Graphic; Source: Bloomberg*

(a) Calculate the change in mean rank order for the 'tech' TNCs shown in Figure 1.18.

> **GUIDANCE**
>
> The 'tech' companies are shown in red in Figure 1.18. You can calculate the mean value for 'tech' companies in 2009 and again in 2018 (note that Facebook only joins this grouping after 2009).

(b) Suggest reasons for the changes shown in the rank ordering of 'tech' TNCs.

> **GUIDANCE**
>
> This question provides you with an opportunity to apply your knowledge and understanding of the growing importance of technology TNCs within global systems. These companies have refined their products and grown enormously in market value over time. Increasingly, most kinds of human activity depend on ICT. The change in rank ordering shown in Figure 1.18 may also reflect the declining importance of some other industrial sectors, such as traditional high-street retailers. Arguably, their decline is linked with the rise of tech companies (and online shopping) as part of a positive feedback loop.

(c) Assess the strengths and weaknesses of this presentation method.

> **GUIDANCE**
>
> This question requires you to weigh up positive and negative traits of Figure 1.18 before giving a final judgement about whether you feel it is, on balance, the best way to display the data. One approach might be to write about how well Figure 1.18 'tells a story' – in other words, does it communicate trends and patterns effectively? Or is it perhaps too complex, meaning that readers 'can't see the wood for the trees'? Another approach when critiquing graphs and charts is to ask: what am I *not* being told that it would be good to also know about? For example, although we are shown what the rankings of different companies are, we lack insight into their actual changing *value*. ExxonMobil looks to have performed badly because it has fallen from first to ninth place. However, perhaps its market value has actually *increased* over time but it just happens that eight other companies have performed even better.

#  Evaluating the issue

▶ *Discussing the importance of technology and TNCs as causes of globalisation*

## Identifying possible contexts, criteria and evidence for the discussion

This chapter's closing debate deals with causality. The focus is the importance of two major global forces – technology and TNCs – as drivers of accelerating globalisation. The question asked here is a big one, however. Before answering, it is important to 'unpack' the three elements under discussion – technology, TNCs and globalisation – in order to gain some analytical structure.

● We can start by thinking critically about the different types of technology that might be discussed, ranging from road transport to the latest advances in science, such as: the newest generation of phones; artificial intelligence; 3D printing; or quantum computing. Different technologies support different globalising processes and offer varying possibilities for global connectivity.

- Think critically about the different kinds of TNC that exist, also. They vary greatly in scale (from businesses operating in just two or three countries to huge global corporations such as Walmart or General Motors). TNCs differ in terms of which sector(s) of industry they operate within, ranging from agriculture and energy to banking and medical research. The largest companies are conglomerates whose activities span multiple industrial sectors. Large agribusinesses, for example, grow and process food (primary- and secondary-sector activities); they also sell their products directly to customers while constantly inventing new ones (tertiary- and quaternary-sector activities). Table 1.5 shows this wide range of economic

activity. Clearly, different categories of TNC give rise to varying kinds of global flow and, as a result, may have more or less significant roles to play in the story of globalisation.

- An analytical framework is required for writing about globalisation. Two approaches were suggested earlier in this chapter (see page 7):
  1 On the one hand, we could discuss the importance of technology and TNCs for different economic, social, cultural and political strands of globalisation.
  2 Alternatively, we might choose to discuss the way technology and TNCs have fostered the growth of different global flows (flows of money, people, merchandise, services and ideas).

Finally, any debate about causality invites consideration of the way cause and effect relationships can become complex. In any system, different processes are often interrelated and work together in ways

| Primary sector | Commercial agriculture is a major global employment provider dominated by TNC giants such as Del Monte and Cargill. Some large UK supermarkets send automated emails to their Kenyan vegetable suppliers requesting increased production quotas whenever store tills register especially good sales for any particular item. This 'just-in-time' ordering is made possible by ICT and global information flows. |
| --- | --- |
| Secondary sector | Manufacturing companies are household names from Samsung and Sony to BMW and Unilever. These companies have extensive production networks. Companies like Samsung are also constantly innovating and advertising new products. It is thus oversimplifying matters to describe them as purely secondary-sector operators. |
| Tertiary sector | Many banks and media corporations have large global networks with specialist branches operating in different world regions. Since 2001, US production company Sony Picture Television has licensed the rights to the *Dragons' Den* franchise, which originated in Japan, to around 30 countries, sometimes taking the step of re-naming the show (in Colombia it is titled *Shark Tank*; in Kenya it's called *Lions' Den*). |
| Quaternary sector | Research and development of new information technology, biotechnology and medical science comes under the umbrella of quaternary-sector work. In China, US firm Intel employs 1000 research staff on the outskirts of Shanghai. ABB, the Swedish-Swiss engineering group, has a research wing in Beijing, as do Nokia and Vodafone. Medical research is increasingly conducted in India by Western firms such as Pfizer and GlaxoSmithKline. |

▲ **Table 1.5** TNCs can be found operating in every sector of industry

which amplify one another's effects. As we shall see, the *interrelations* between technology and TNCs have an important bearing on globalisation patterns and trends. It is not simply a question of deciding which force is most important.

## View 1: Technology is the most important cause of globalisation

Technology has played a critically important role in the way the world's many peoples, places and environments have increasingly become joined together as part of a single global system. Over many millennia, ever-denser trade, migratory and communication networks have evolved.

In the 1800s, railways and steam engines helped to connect countries and continents. These and other technologies such as the telegraph were instrumental in allowing European countries to build global empires which existed until the 1960s and 1970s. During the last few decades – viewed by many as a 'golden era' of globalisation – container vessels and communication networks have played a vitally important supporting role. Without these technologies, the **global shift** of heavy and light industry could not have happened. China's emergence in the 1980s and 1990s as the new 'workshop of the world' was reliant on container vessels exporting vast quantities of merchandise to developed and other emerging markets alike (see Figure 1.19).

▲ **Figure 1.19** Container vessels, which carry merchandise around the world, have sometimes been called 'globalisation's lifeblood'

▲ **Figure 1.20** The former First Lady of the United States, Michelle Obama, participated in the #BringBackOurGirls campaign which raised global awareness of the kidnapping of 276 schoolgirls in Chibok, Nigeria

Since the 1990s, technology has supported exponential growth in flows of data, along with the movement of unprecedented numbers of people.

- In 2017, 15 zettabytes of data were transmitted through global data centres. This is an unfathomably large amount of information. It is equivalent to the memory contents of *trillions* of phones – think of just how many films, images, conversations and bits of data this represents.
- Social causes can go viral in an instant: the success of awareness-raising campaigns like *#Kony2012* and *#BringBackOurGirls* demonstrates how technology facilitates the rapid global spread of ideas (see Figure 1.20).
- ICT also fosters cultural globalisation – defined here as the growth of a global culture with shared traits and social norms. Released in 2012, Psy's *Gangnam Style* became the first music video to achieve more than 1 billion views online. The YouTube RSS feeds of South Korea's large diaspora population helped the song diffuse globally; a few tweets from celebrities including Britney Spears and Tom Cruise ensured massive exposure.
- More than one-quarter of a billion people now live in countries they were not born in. Technology can be viewed as a cause of

increased flows of people around the world, a movement which in turn generates other kinds of global flow. The remittance flows generated by economic migrants contribute US$500 billion annually to global economic systems; migrants also transport their own ideas and culture when they travel. But without different forms of transport, none of this would happen.

- Increasingly, migration is also supported by digital technology. In 2015, a Facebook group called Stations of the Forced Wanderers helped over 100,000 refugees to exchange advice on how best to avoid authorities and find routes across European borders using GPS information.
- Online media representations of places may also affect people's initial decision to migrate. If YouTube films portray a particular country positively, this could prompt viewers in other countries to relocate there.
- Migration may become psychologically easier when people can maintain long-distance social relationships online. Free Skype and Messenger services allow migrants to maintain strong links with family and friends they have left behind.

## View 2: TNCs are the main cause of globalisation

TNCs have played a pivotal role in the growth of globalisation. The largest manufacturing and retail businesses have built extensive global production and consumption networks (see Figure 1.21). They employ various investment strategies to expand their operations. These range from establishing new low-cost factories and offices in emerging economies to bespoke glocalisation strategies which help companies endear themselves to consumers in widely differing local contexts.

These companies are not only responsible for economic globalisation. They also bring social and cultural changes to global systems.

▲ **Figure 1.21** Well-known global brands show their presence on a high street in Manchester

- The arrival of companies like McDonald's and Yum! Brands (KFC) may bring about change in the cultural preferences of a country's population. In particular, food companies have some responsibility for introducing meat and dairy products to countries where diets were traditionally vegetarian. According to the UN Food and Agriculture Organization, globalisation has led to a dramatic shift of Asian diets away from staples (such as rice) and increasingly towards livestock and dairy products, vegetables and fruit, and fats and oils.
- More subtle global cultural changes have resulted from the runaway success of US media and technology companies like Apple and Google. Their operating systems and software have contributed to the global spread of not only the English language but also European and North American cultural traditions. For example, traditional Western festivals and events – including Christmas, Halloween and Valentine's Day – appear on iPhone calendars apps. Global media companies, including the BBC, Disney and MTV, also mark these times of year in their programming output. Notably, there has been some resistance against this in Pakistan – in 2016, President Mamnoon Hussain declared that Valentine's Day is 'a Western cultural import' that threatens Pakistani values (see page 70).

It is not only manufacturing and technology/media companies that have played an important role in globalisation. The world's largest food, mining and energy companies have great influence too. Some agribusinesses, like US TNC Cargill, have globalised their operations in ways which have transformed the lives of rural societies in developing and emerging countries. Cash crops such as soya are grown on an ever-greater scale to feed rising consumer demand in global markets; energy companies continue to scan the planet for fossil-fuel resources to exploit.

- Rural people are sometimes left landless because of the so-called land grabs that may result from TNC activities (see page 158).
- Consequently, millions of rural migrants have arrived in African, Asian and Latin American cities where they are far more likely to encounter global cultural influences or may even join a workforce which produces branded commodities for TNCs.

Finally, we must remember the important global role of TNCs from the financial sector. This category includes banks, brokers, accountancy, consultancy, investment and law firms. One view is that these companies have the greatest influence of all over globalisation because of the way world governments consistently cater to their desire for free-market policies and 'frictionless' trade in money markets. Critics say that the movements of capital channelled around the planet by financial-sector TNCs have brought disproportionate benefits to an already wealthy global elite which has a vested interest in perpetuating 'business as usual' globalisation (see page 138).

## View 3: Technology and TNCs are equally important because of their interrelations

When we inquire about the relative importance of technology and TNCs, perhaps we are asking the wrong question. This is because the two entities are so deeply interrelated that they are often indistinguishable. For example, do you think of Google and Microsoft as technologies or businesses? An iPhone is a piece of technology, to be sure, but it is also the signature product of Apple, the world's largest TNC by market value in 2018 and the first company to ever be valued at more than 1 trillion dollars. The second company to pass the same valuation milestone was Amazon, another example of a company that also creates technologies which further accelerate globalisation.

David Harvey has written at length about how global capitalism depends on continued cycles of innovation driven by businesses in search of greater profit. Evidence supporting this argument is all around us: most of the technologies on which globalisation depends were developed by TNCs including IBM, Microsoft and, more recently, the FANGs. In developed economies, where markets quickly become saturated, the research priority for these companies has long been to devise ever-more superior products. Moreover, part of the logic behind creating spatial divisions of labour (see page 28) is to reduce production costs so that more profit can be reinvested in research and development of new products. Firms like Apple make great use too of advertisements that aim to convince existing customers they should throw away their current (fully working) phone because it lacks features the latest models now possess.

In short, when profits begin to fall, the stimulus arises for these TNCs to invest heavily in new technologies. They do so for fear of losing their position as top-ranked businesses (see Figure 1.18). In turn, fresh innovation in phones and desktop devices may create a need for faster data download speeds. This prompts the suppliers of fibre-optic cables and mobile data companies to improve their own technologies. The result is a continued cycle of innovation driven by competing hardware, software and infrastructure companies. Sometimes, this cycle

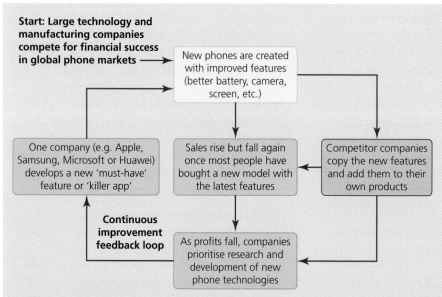

**Start: Large technology and manufacturing companies compete for financial success in global phone markets** → New phones are created with improved features (better battery, camera, screen, etc.)

One company (e.g. Apple, Samsung, Microsoft or Huawei) develops a new 'must-have' feature or 'killer app'

Sales rise but fall again once most people have bought a new model with the latest features

Competitor companies copy the new features and add them to their own products

**Continuous improvement feedback loop**

As profits fall, companies prioritise research and development of new phone technologies

▲ **Figure 1.22** Positive feedback in global systems explains why phones have made enormous 'jumps' forward in their capabilities in recent years

is accelerated further by TNC mergers and acquisitions (see page 27). Google purchased Android; Microsoft bought LinkedIn; Facebook owns Instagram.

This constant technological innovation loop – driven by businesses and in turn driving globalisation – is a form of **positive feedback** (see Figure 1.22). Empirical evidence for this is provided by Moore's Law, a proposition stating that computer chip performance doubles in power every two years. Between 1970 and around 2010, computing speeds grew in line with this rule (more recently it appears some fundamental technological limits may have finally been reached). The result was exponential growth in the capabilities of computers, tablets and more recently phones. Figure 1.23 illustrates this relationship (the graph uses a logarithmic *y*-axis and so shows a straight rather than curved line).

In summary, enormous gains in computer power helped propel global systems into the twenty-first century – but they were the result of TNCs such

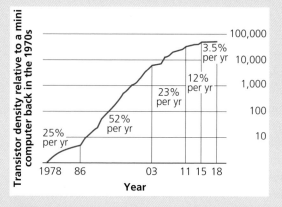

▲ **Figure 1.23** A semi-logarithmic graph showing Moore's Law (the processing power available to computers –determined by transistor density – doubled every two years between around 1980 and the early 2000s). *FT Graphic; Sources: Computer Architecture by John Hennessy and David Patterson; Bloomberg*

as Microsoft, Apple and their many partners constantly seeking out a business edge to increase the financial returns from their global investments.

# Reaching an evidenced conclusion

How far should either technology or TNCs be held most responsible for the acceleration of globalisation in recent decades? This discussion has shown there is no simple conclusion to such a question. Different types of technology or category of TNC may each be seen as the cause of a specific aspect of globalisation (but not global system growth in its entirety). We could argue, for example, that transport innovation is the driving force which has unlocked people's potential to migrate, resulting in *social and cultural globalisation* on a global scale – and TNCs have only a supporting role to play in this particular narrative. Or we might view TNCs as the driving force behind *economic globalisation* – because of how they knit together the vast global-scale supply networks which provide people in many countries with their everyday foods, goods and services.

As we have seen, a strong case can also be made that TNCs and technology have *jointly* helped shape globalisation in recent decades. It is not just a matter of concluding that 'both are important' though. Rather, they are so closely interrelated that in essence they form two sides of the same coin. Increasingly, TNCs even develop and pay for the fibre-optic cables upon which global connectivity now depends (as Figure 1.13 showed). The social forces that shape technological innovations are often the same as those driving the expansion of TNCs into new markets. Both are led by people striving for an advantage in a highly competitive global economy; and both result in further rounds of globalisation.

 **KEY TERMS**

**Global shift** The international relocation of different types of industrial activity, especially manufacturing industries. Since the 1960s, some economic activities have all but vanished from Europe and North America. In turn, the same activities now thrive in Asia, South America and, increasingly, Africa. The term is widely associated with the work of geographer Peter Dicken.

**Land grab** The acquisition of large areas of land in developing countries by domestic forces or international investors, governments and sovereign wealth funds. Indigenous people who have occupied land for centuries or millennia may be told they no longer have the right to remain where they have always lived.

**Elite** A group of people who are economically and/or socially powerful. Many are highly skilled professionals, for example with expertise in finance, sport or the arts. The hyper-wealthy elite's money may be inherited or entrepreneurial in origin. Elite migrants may move to new countries for work reasons or perhaps to shelter their wealth from excessive taxation in other states.

**FANGs** An acronym for four hugely profitable technology TNCS: Facebook, Amazon, Netflix and Google.

**Positive Feedback** When change in a system triggers further 'knock-on' or 'snowballing' changes. As a result, system changes accelerate in potentially unmanageable ways.

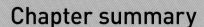

# Chapter summary

✔ All of the world's people, places and environments can be thought of as being linked together by flows of money, people, ideas, information, services and merchandise (goods). Over time, global systems have grown in size and complexity.

✔ Globalisation is an umbrella term for a variety of processes of change which have triggered a step change in global system growth in recent decades. Key changes include the way global flows have lengthened, sped up and now push deeper into societies and places than in the past.

✔ Globalisation is a contested idea that can be defined and analysed in many ways. One approach is to study it as a multi-strand process with economic, social, cultural and political dimensions. Views differ about the costs and benefits of all aspects of globalisation.

✔ Improvements in transport and technology over time have led to time–space compression and the shrinking-world effect. Global data flows have transformed the way people work, think and communicate at a global scale.

✔ Trade and TNCs are central to the growth of global systems and globalisation. Spanning all sectors of industry, TNCs link places together through their networks of production, distribution and consumption. Flows of capital, merchandise and services are directed around the world by TNCs; flows of ideas (including values and culture) are an important by-product of this.

✔ Working together in tandem, technology and TNCs are dual forces that have reshaped the world in recent decades. Large technology companies including Apple, Google, Facebook and Amazon are at the forefront of this latest global industrial revolution.

# Refresher questions

1 What is meant by the following geographical terms? Global flow; global norm; hyperconnected.

2 Using examples, explain how modern globalisation differs from the earlier phases of world development.

3 Explain why some people view globalisation positively while others have a more negative perspective.

4 Using examples, outline the meaning of: economic globalisation; social globalisation; cultural globalisation; and political globalisation.

5 Outline the strengths and weaknesses of the KOF Globalisation Index.

6 Using examples, explain how the following technologies have contributed to global system growth: container shipping; air travel; fibre-optic cable; social media services.

7 What is meant by the following geographical terms? Global capitalism; offshoring; outsourcing; joint venture; glocalisation.

8 Using examples, outline reasons why some TNCs make use of a spatial division of labour in their operations.

9 Explain why TNCs in the technology sector, such as the FANGs, have become key players in global systems.

# Discussion activities

1   Individually, draw a mind map to answer the question: 'How globalised am I'? This deceptively simple task requires an analytical framework. You might think about how economically or culturally globalised you are, for instance. Or you might consider the ways you have participated in different global flows – what, if any, involvement have you had with global flows of people, ideas and merchandise? Aim to write roughly one side of A4 paper. Work together in small groups and compare your accounts upon completion.

2   In groups, discuss the extent to which people of your own age group enjoy a more globalised lifestyle than those of your grandparents or other older people you know. Think about your exposure to different cultures and ideas, for example. What explains any differences that you have identified between age groups?

3   In groups, write a definition of globalisation (without making reference to anything written in this book or elsewhere). Next, compare what you have written with other people, and also with the definitions in Figure 1.3 (see page 5). What did you include and what did you leave out (that you could have included) in the definition you wrote?

4   In pairs, research the geography of a TNC you are interested in. This could be a media company, a food producer, an oil company, a bank or something else entirely. Try to find details of the production and consumption networks for your chosen company (the countries where goods and services are made, and the countries where they are sold and consumed). Look for evidence of strategies the TNC uses, such as joint ventures or glocalisation. When each pair of students has completed their research, findings can be shared with the class, possibly as part of a formal presentation.

# FIELDWORK FOCUS

The topic of globalisation provides interesting opportunities to devise a creative independent investigation. Anyone wanting to carry out this kind of work will need to think critically about the best way to measure globalisation (which is not necessarily a straightforward thing to do).

A   *Comparing two different places to see how globalised they are.* First, you must devise an analytical framework to help you understand how 'globalised' your chosen places are: identify particular measurable criteria that represent economic globalisation, social globalisation, cultural globalisation etc. For example, you might want to look at the proportion of international retailers (TNCs) who have a high-street presence. You might also search for secondary data showing the number of people in each area believed to

have migrated there from abroad. A study of cultural glocalisation could involve asking passers-by to fill in a questionnaire about the food they eat, or the TV shows they watch, in order to try to quantify how 'global' their cultural tastes and preferences are. There are many challenges involved in devising an ambitious study like this, but the results could be very rewarding and result in a high mark being awarded.

B   *Interviewing selected people of different ages or ethnicities in order to investigate the extent to which contrasting groups of people participate in global systems.* This type of study uses a stratified sampling technique: you might decide to interview a target number of people aged under 30 and another group aged over 70, for example. The questionnaire could focus on people's travel experiences or their use of social media.

# Further reading

Castree, N., Coe, N., Ward, K.G. and Samers, M. (2003) *Spaces of Work: Global Capitalism and Geographies of Labour*. London: Sage.

Dicken, P. (2014) *Global shift: mapping the changing contours of the world economy*. 7th ed. London: Sage.

Harvey, D. (1989) *The Condition of Postmodernity*. Oxford: Blackwell.

Held, D., McGrew, A., Goldblatt, D. and Perraton, J. (1999) *Global Transformations: Politics, Economics and Culture*. Cambridge: Polity Press.

Martin, R. (2004) Geography: making a difference in a globalizing world. *Transactions of the Institute of British Geographers*, 29(2), 147–150.

Massey, D. (1993) Politics and space/time. In: M. Keith and S. Pile, eds. *Place and the Politics of Identity*. London: Routledge.

Stiglitz, J. (2002) *Globalisation and its Discontents*. London: Allen Lane.

Waters, M. (2000) *Globalization*. 2nd ed. London: Routledge.

# Power relations in global systems

People and businesses are not free to roam around the world at will. National governments must first agree to participate in globalisation by allowing different global flows to cross their borders. This chapter:

- explains the strategies used by national governments to 'switch on' local places and people to global flows
- investigates supranational influences on global trade, the free movement of people and flows of ideas
- analyses the varied ways in which powerful states influence how global systems work to their own advantage
- evaluates the extent to which globalisation involves the imposition of Western ideas on the rest of the world.

## KEY CONCEPTS

**Power** The ability to influence others by purposely affecting change or maintaining equilibrium. Power is vested in citizens, governments, institutions and other players at different geographical scales. Equity and security, both environmental and economic, may be gained or lost as a result of how powerful forces operate.

**Scale** An important organising concept for geography: structures, flows, places and environments can be identified at a variety of geographic scales, including local areas and national (state) territories.

**Superpower state** A country that projects its power and influence at a truly global scale.

**Supranational** A geographic scale that transcends national boundaries. Supranational organisations and agreements may have powers that surpass or increase the influence of national governments.

 # National governments and global flows

▶ *How and why have national governments allowed some global flows to operate more easily?*

## Free trade and investment policies

The degree to which different countries are integrated into global financial and market systems varies greatly. Some countries are far more open to free trade and investment flows than others because of the actions and attitudes

▲ **Figure 2.1** In a game of Monopoly, some players tear around the board buying whatever properties they land on. However, in the real world, global investors have not always been allowed to enter local markets because of government policies and laws. In recent years, barriers have been rising in some parts of the world

of their national governments. Real-world political barriers to global flows – and there are many – can feel decidedly at odds with 'shrinking-world' rhetoric (see page 18):

- Some of the views about globalisation we encountered on page 5 seemed to imply a borderless, frictionless globe over whose surface companies can roam freely like players on a Monopoly board, buying up properties with each roll of the dice (see Figure 2.1).
- This kind of hyperglobal view of the planet gained traction around the turn of the millennium; in the 1990s, Francis Fukuyama's 'End of History' thesis argued that the global spread of Western ideas and neoliberal free-market norms (see page 8) represented a 'finished project' for world development.
- But the world today increasingly feels very different. Barriers to trade and movements of people are rising, not falling, in some parts of the world. Later in this book, Chapter 6 takes a critical look at the extent to which recent 'global shocks' have challenged the idea that globalisation is an unstoppable force.

This section focuses on how national governments *helped* global trade and investment flows across their borders to grow greatly in size between the 1980s and early 2000s (see Figure 2.2). A range of neoliberal policies, laws and economic tools have been deployed in different places and at different times with the aim of attracting 'footloose' global investors in a competitive global market place. These policies include the establishment of free-trade zones, low-tax regimes and 'light touch' regulation of foreign direct investment. Writing in the 1990s, Manuel Castells analysed the strategic use of these tools by governments to ensure their states become 'switched-on' to parts of the global economy and its 'network society'.

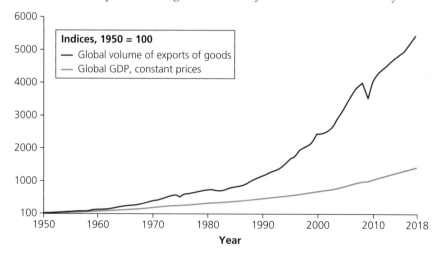

▲ **Figure 2.2** The indexed data in this chart (1950 value = 100) show the aggregate global volume of exported goods growing at a near-exponential rate between 1950 and the early 2000s. This was due in part to positive national government attitudes towards participation in global trade flows

## Neoliberalism, deregulation and privatisation

Neoliberalism – also known as free-market liberalisation – is a political and economic philosophy. It is strongly associated with the policies of US President Ronald Reagan and Margaret Thatcher's UK government during the 1980s; subsequently, neoliberal thought was adopted by policy-makers in other countries too. Essentially, neoliberalism follows two simple beliefs:

1 Government intervention in markets – including global markets and financial flows – impedes a country's economic development.
2 As overall wealth increases, trickle-down will take place from the richest members of society to the poorest.

In the UK, neoliberalism meant restrictions being lifted on the way companies and banks operated. For example, the deregulation of the City of London in 1986 removed large amounts of 'red tape' and paved the way for London to become a leading global centre for financial services and the home of many super-wealthy 'non-dom' billionaires (see Figure 2.3, also page 155). Successive UK governments also encouraged foreign investors to compete for a stake in privatised national services and infrastructure. Until the 1980s, important assets, such as the railways and energy supplies, were owned by the state. However, running these services often proved costly: they were therefore sold to private investors in order to reduce government spending and to raise money. Over time, ownership of many assets has passed overseas. For instance, the French company Keolis owns a large stake in southern England's railway network and the energy company EDF is owned by Électricité de France. In recent years, the UK government has approached Chinese and Middle Eastern investors to help fund new infrastructure projects on numerous occasions.

## Establishing free-trade zones

One very important reason for the acceleration of globalisation after the 1970s was changing government attitudes in regions outside of Europe and North America. Notably, Asia's three most populated countries – China, India and Indonesia – all embraced trading to global markets as a means of meeting economic development goals. In each case, the establishment of special economic zones (SEZs), government subsidies and changing attitudes to FDI played important roles.

- In 1965, India was one of the first countries in Asia to recognise the effectiveness of developing its exports in order to promote growth. Today, there are nearly 200 Indian SEZs.
- Before 1978, China was a poor and politically isolated country, 'switched-off' from the global economy. This changed when Deng Xiaoping began the radical 'Open Door' reforms which allowed China to embrace globalisation while remaining under one-party authoritarian rule. Coastal SEZs were crucial to the new plan – many of the world's largest TNCs were quick to establish offshore branch plants or build

Pearson Edexcel   AQA   OCR   WJEC/Eduqas

### KEY TERMS

**Trickle-down** The positive impacts on poorer, peripheral regions (and people) caused by the creation and accumulation of new capital in wealthier core regions (and their societies). The phrase is also associated with neoliberal arguments in favour of lower taxation rates (on the basis that it leaves the rich with more money to invest in ways which benefit the poor).

**Special economic zone (SEZ)** An industrial area, often near a coastline, where favourable conditions are created to attract foreign TNCs. These conditions include low tax rates and exemption from tariffs and export duties.

▲ **Figure 2.3** Canary Wharf in London's Docklands is a leading world financial centre: large global flows of investment capital and high-skilled labour make this a highly connected place. UK government decision making played an essential role in its success by creating a liberal investment regime while also assisting with the area's regeneration

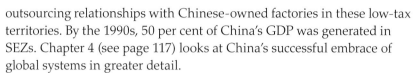

outsourcing relationships with Chinese-owned factories in these low-tax territories. By the 1990s, 50 per cent of China's GDP was generated in SEZs. Chapter 4 (see page 117) looks at China's successful embrace of global systems in greater detail.

- Indonesia provides another striking example of political influences on global interactions (see page 60).

## Low-tax regimes

One way in which the USA, UK, Germany and other developed countries have benefited greatly from globalisation is through the large corporation taxes paid by the many high-value TNCs domiciled within their borders. Apple, which is headquartered in California, paid US$16 billion to the US government in 2017; the top tax payer was ExxonMobil, which handed over US$31 billion.

- In recent years, some European-based TNCs have been enticed to relocate their headquarters to Ireland, Switzerland, Luxemburg or the Netherlands, where corporate taxes are low (in 2019, the UK rate was 19 per cent, around twice that of Switzerland; in the USA, the rate was 35 per cent until President Trump reduced it to 21 per cent in 2017).
- This demonstrates how national governments can actively encourage TNCs to relocate inside their own territories, thereby 'capturing' a greater share of global trade and capital flows – if you refer back to Figure 1.11 on page 16, you will notice how high low-tax Switzerland and the Netherlands score according to the KOF Globalisation Index.

In 2010, petrochemical firm INEOS moved its headquarters from the UK to Switzerland. This corporate migration yielded an estimated saving of almost half a billion pounds over five years. Why don't more TNCs do the same? There are practical reasons explaining the reluctance of many companies – especially the largest household names – to move home. These relate to brand authenticity, corporate responsibility, public perception and security. The last of these is especially important. TNCs sometimes look to the national government where they are headquartered for support during a financial crisis, or when their overseas assets become threatened by conflict or nationalisation.

- The oil company Repsol sought support from its country of origin, Spain, when Argentina's government seized control of Repsol's Argentinian investments in 2012.
- During the global financial crisis (GFC) of 2008–09, General Motors and Chrysler looked to the US government for support, while the Royal Bank of Scotland was 'bailed-out' by the UK treasury.

Instead of relocating their headquarters to a country with lower taxes, many TNCs instead choose the strategy of transfer pricing to reduce their tax burden. This involves routing profits through subsidiary (secondary)

 **KEY TERMS**

**Corporate migration** When a TNC changes its corporate identity, relocating its headquarters to a different country.

**Transfer pricing** A financial flow occurring when one division of a TNC based in one country charges a division of the same firm based in another country for the supply of a product or service. It can lead to less corporation tax being paid.

companies owned by the parent company. These subsidiaries are based in a low-tax state like Ireland or possibly an offshore **tax haven** (see Figure 2.4). Around 40 so-called tax havens offer nil or nominal taxes. Some are sovereign states, such as Monaco. Another, the Cayman Islands, is an overseas territory of the UK that has its own tax-setting powers. It is not just companies who route money in this way. Some wealthy expatriates try to limit their personal tax liability by migrating to a tax haven.

## Laissez-faire economics and FDI

Some national governments adopt a broadly 'laissez faire' approach to FDI. This means stepping back and allowing market forces to operate without too much political fuss and bother. Acting very much in line with this philosophy, successive British governments have displayed a largely relaxed attitude towards FDI, making the UK the world's second biggest market for foreign takeovers. Cadbury was acquired by the US conglomerate Kraft in 2010, for example; since then, Jaguar Land Rover (JLR) (see page 90) has been sold on to India's Tata Motors, while Chinese investors have acquired a large stake in Thames Water, the company that provides the UK's capital city with its most vital need. Recently, however, some UK politicians have argued for greater controls on foreign ownership of strategically important industries. Selling chocolate to an overseas bidder is one thing, they say, but selling off control of our water is another.

> **KEY TERM**
>
> **Tax haven** A country or territory with a nil or low rate of corporation tax.

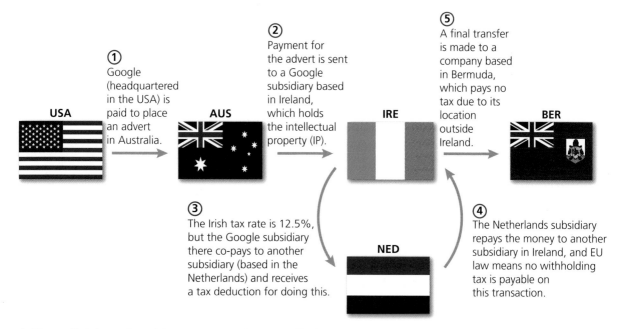

**① Google** (headquartered in the USA) is paid to place an advert in Australia.

**② Payment** for the advert is sent to a Google subsidiary based in Ireland, which holds the intellectual property (IP).

**③** The Irish tax rate is 12.5%, but the Google subsidiary there co-pays to another subsidiary (based in the Netherlands) and receives a tax deduction for doing this.

**④** The Netherlands subsidiary repays the money to another subsidiary in Ireland, and EU law means no withholding tax is payable on this transaction.

**⑤** A final transfer is made to a company based in Bermuda, which pays no tax due to its location outside Ireland.

USA    AUS    IRE    BER    NED

▲ **Figure 2.4** A transfer pricing strategy used by some TNCs to reduce their corporate tax bills. As a result, huge capital flows cross the borders of countries whose governments have opted for very low rates of corporation tax

In contrast with the UK's 'light touch' investment regime, governments in other countries do more to scrutinise and regulate corporate acquisitions and mergers (see page 27) involving overseas investors. Larger-scale organisations such as the EU can sometimes have a say in what happens too (the EU has a legal duty to prevent powerful monopolies from emerging). As a result, there are plenty of political and legal obstacles for TNCs to remain mindful of when going about their global business in some local contexts.

- The French and Italian governments monitor, and potentially halt, unwanted foreign takeovers in sectors deemed 'strategically important', such as energy, defence, telecoms and food.
- The Committee on Foreign Investment in the United States closely scrutinises inbound foreign takeovers, while the Canadian and Australian governments have powers to intervene. The US government permits the use of 'poison pill' strategies by well-established American-based companies in order to stave off unwanted attention by potential foreign buyers. When threatened by a hostile takeover, firms can dilute the value of their shares, thus weakening the voting power of a potential acquirer.
- India requires that TNCs work in partnership with Indian companies (though rules were recently relaxed for 'single-brand' retailers like IKEA).
- China's government blocked Coca-Cola's acquisition of Huiyuan Juice in 2008 (although it recently allowed British TNC Diageo to buy Shui Jing Fang, a famous brand of spirits).

Should the UK government do more to prevent control of its own companies being lost overseas, as some other governments do? The case against making any changes to investment law is that the UK remains the world's fifth largest economy measured in terms of output. In part, this success may be attributable to unimpeded high inward capital flows from US, Japanese, German and Chinese firms, among many others.

### Non-participation in global flows

Not all state governments choose to participate fully in global systems. A minority have 'opted out' of global systems in various ways, viewing neoliberalism as a form of hegemony that should be resisted. Although neoliberal ideas may have become widely accepted as 'common sense' over time, critics of free-market global systems say globalisation has increased inequality, injustice and conflict on a worldwide scale (themes which are investigated in Chapters 4 and 5). Since 2000, Bolivia's Evo Morales, Zimbabwe's Robert Mugabe and North Korea's Kim Jong-un have all made attempts to resist globalisation (see Figure 2.5).

Before his death in 2013, Venezuela's leader, Hugo Chávez, called repeatedly for other Latin American and Caribbean countries to resist the 'free-market neoliberal order' promoted by the USA and the Washington-based World Bank and IMF. He rejected IMF and World Bank offers of support for Venezuela while taking back control of oil operations inside the

 **KEY TERM**

**Hegemony** When less powerful people (or countries) are convinced to accept ideas, beliefs or policies that are not in their best interests by more powerful people (or countries). The latter achieve this using a range of persuasive 'soft power' strategies including diplomacy, aid and domination of media and educational institutions.

country which TNCs like Chevron, ExxonMobil and BP had invested in. In 2004, Chávez called for more social organisations and institutions around the world to work together 'to build alternative models of development in the face in globalisation'.

However, those governments who criticise global capitalism have yet to persuade the rest of the world that a viable alternative model for state building exists.

- Life expectancy in North Korea, which has effectively 'opted out' of globalisation (although there are early signs this may be changing), is more than ten years lower than in neighbouring South Korea.
- Zimbabwe has only recently recovered from an episode of hyperinflation (when economic mismanagement led to interest rates reaching 79 billion per cent).
- For a while, Venezuela managed to successfully fund a non-capitalist development path thanks to its large oil reserves and revenues. However, falling oil prices in 2015–16 crippled Venezuela's development model, leading to devaluation of its currency, the onset of hyperinflation and mass out-migration. In early 2019, the UN estimated 3 million Venezuelans (ten per cent of the population) had fled the country.

On a less extreme scale, particular types of trade flow may be disallowed by national governments for reasons ranging from political disagreement to biodiversity threats. Figure 2.6 shows how some countries have selectively 'opted out' of participating in certain kinds of trade.

## Migration policies

Alongside trade policies, national governments take strategic decisions to enable or prevent flows of people crossing their borders. States vary greatly in terms of (i) the size of their population that is comprised of economic migrants and (ii) the rules which govern immigration (see Table 2.1). Differing domestic needs and political priorities drive decision-making; so too do broad differences in levels of economic engagement with the rest of the world.

- Host countries differ greatly in terms of how liberal their international migration rules are. Laws governing economic migration vary over time in line with changes in workforce needs. The UK government adopted a broadly 'open door' approach to international migration in the 1950s and again during the early 2000s. Both decisions related in part to skills and labour shortages arising during those historical periods.
- In order for a country to become deeply integrated into global systems, its government may need in any case to adopt relatively liberal immigration rules. Inward investment from TNCs may depend in part on the ease with which a company can transfer senior staff into a particular nation. For example, many of the world's leading law firms have regional offices spanning the globe, from Singapore to Moscow. In

▲ **Figure 2.5** Kim Jong-un is often portrayed as being the leader of a 'rogue state'. Supporters of his regime view North Korean as a rare and brave example of a state which has tried to resist Western hegemony

**Prohibited flows**

**Cuba** 🚫 **USA (until 2015)**

The USA imposed a **complete trade embargo** on communist Cuba in 1962 as a result of Cold War antagonism between the two countries. The result? A commercial and financial blockade.

**World** 🚫 **China**

Not all **information flows** are allowed to enter China. For instance, internet users there are not allowed access to the BBC's Chinese-language website service.

**Australia** 🚫 **New Zealand**

For 50 years, imports of Australian **honey** were banned in New Zealand for fears of a 'bio-security threat' (Australia's bees suffer from a disease that New Zealand beekeepers have been keen to avoid).

**China** 🚫 **Europe**

In 2005, the EU briefly banned the further imports of **cheap Chinese textiles** — especially women's bras — in an attempt to protect its own manufacturers. This was dubbed 'bra wars' by the media.

▲ **Figure 2.6** Prohibited global flows

order to maintain their global networks, these companies depend on foreign states granting their staff permission to relocate permanently to overseas offices.

- Top lawyers belong to a wider 'global elite' of professionals and high-wealth individuals. Such people are likely to encounter fewer obstacles to international migration than lower-skilled migrants. Their talent and wealth makes them more likely to qualify for a visa or residency, especially in states where a points-based immigration system exists, such as Australia. Some elite migrants live as 'global citizens' and have multiple homes in different countries (see page 126).

Few countries have laws preventing the out-movement of people because this contravenes the United Nations Universal Declaration of Human Rights (UDHR); Article 13 guarantees: 'Everyone has the right to leave any country, including his own, and to return to his country.' As a result, almost all states are, in theory, potential source countries for unlimited out-migration.

- One notable exceptions to this rule is North Korea, whose government still requires that its citizens obtain an exit visa before being allowed to leave.
- In the past, citizens of the Soviet Union faced similar restrictions on their freedom of movement.
- In Saudi Arabia and Qatar, some foreign migrants must apply for an exit visa before being allowed to go home.

| Japan (stricter migration rules) | Less than two per cent of the Japanese population is foreign or foreign-born. Despite the growing status of Japan as a major global hub from the 1960s onwards, migration rules have made it tough for newcomers to settle in the country permanently. Nationality law makes the acquisition of Japanese citizenship by resident foreigners an elusive goal (the long-term 'pass-or-go-home' test has a success rate of less than one per cent). Japan faces the challenge of an ageing population, however. There will be three workers per two retirees by 2060. Many people think that Japan's government will need to loosen its grip on immigration to solve this conundrum. |
|---|---|
| Australia (stricter migration rules) | While Singapore has a high percentage of foreign workers, the proportion found in Australia is lower due to a recent history of restrictive migration policies. The country currently operates a points system for economic migrants called the Migration Program. In 2017, only 245,000 economic migrants were granted access to Australia (this figure included the dependants of skilled foreign workers already living there). The top five source countries were India, China, the UK, the Philippines and Pakistan. Until 1973, Australia's government selected migrants largely on a racial and ethnic basis. This was sometimes called the 'White Australia' policy. |
| Singapore (liberal migration rules) | Until recently, Singapore was rated as an emerging economy. Now a developed nation, this city-state is unusual in many respects. Among its 5 million people there is great ethnic diversity due to its past as a British colonial port and subsequent transformation into the world's fourth-largest financial centre. Plenty of global businesses and institutions have located their Asia-Pacific head offices in Singapore, including Credit Suisse and International Baccalaureate. Many foreign workers and their families have relocated there and, as a result, Singapore has many international schools. |

▲ **Table 2.1** Japan, Australia and Singapore have differing attitudes towards immigration

# Data flow policies

National government attitudes towards the free movement of data is an interesting and highly topical aspect of global systems for geographers to engage with. What makes this area of governance especially challenging for policy-makers is the way rapid rates of innovation (see page 36) often outstrip the speed with which lawmakers can develop an adequate regulatory framework for newly-arrived technologies. For instance, the use of social media to radicalise young people or undermine democracies (see pages 73–74) has become a pressing concern for many governments, yet attempts to tackle these issues often meet with abject failure due to the way increasingly powerful new technologies, platforms and apps continue to be made available to ever-larger numbers of people.

In some countries, data flows are blocked or censored by the state. On the face of it, this is surprising, given that unfettered access to ICT may stimulate outsourcing, supply-chain growth and sales of digital services – all of which can be profitable activities. However, the social interactions between citizens that happen online are viewed as a threat by some authoritarian and non-democratic governments. This can lead to restrictions on social networking, resulting in two kinds of switched-off society.

- *Disconnected states.* Some states limit their citizens' access to cross-border flows of information, resulting in a **splinternet**. Facebook, Twitter and YouTube remain unavailable to Chinese users as part of the 'Great Firewall of China' (in a parallel example of cultural isolationism, only 34 foreign films are allowed to be screened at Chinese cinemas each year). Google withdrew from China in 2010 due to the Chinese government's insistence that search engine results should be censored. Yet, while there is little external connectivity, 800 million Chinese citizens (2018 data) freely interact with one another within a cyberspace 'walled garden' using local blog sites, such as Youku (see Figure 2.7). Other states with similar restrictions include Iran and Pakistan.
- *Disconnected citizens.* In some states, people additionally lack the means to communicate digitally with their fellow citizens within national boundaries. Although cost is of course a factor, the fact that 25 million North Koreans largely have no access to the internet at all is a result of political decision-making. In the past, the authorities in Saudi Arabia have restricted messaging using BlackBerry because security forces could not crack the BlackBerry encryption code and were therefore unable to eavesdrop on private conversations. One source of this paranoia was the use of BlackBerry devices by the 2008 Mumbai terror attackers, which led to calls for a ban in India, too.

**KEY TERM**

**Splinternet** A global internet that is increasingly fragmented (or 'Balkanised') due to nation-states filtering content or blocking it entirely for domestic or international political purposes.

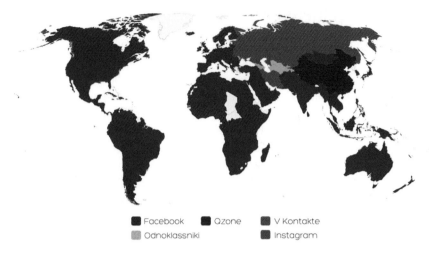

Facebook  Qzone  V Kontakte
Odnoklassniki  Instagram

▲ **Figure 2.7** World map showing each country's most popular social networking site in 2018. From a global perspective, China is a 'walled garden' whose citizens communicate with one another using several platforms, the most popular of which is Qzone (700 million users, 2018)

Restrictions on use therefore operate at two geographical scales: the national and the personal. Worldwide, around 40 national governments have one or both types of limit in place.

## ② Supranational influences on global systems

▶ *How are global flows enabled and directed by the work of intergovernmental organisations (IGOs), world-region trade blocs and other global groups?*

This section deals with the supranational political and economic frameworks which global systems both depend on and give rise to. Decisions about trade, migration and data flows taken by individual national governments are often embedded within much larger-scale politico-economic contexts and legal structures, including numerous treaty-based **intergovernmental organisations (IGOs)** and world-region **trade blocs**. For example, the individual policies adopted by each EU member-state government must always be aligned with the broader rules they have all agreed to as a condition of group membership. At a larger scale yet, the majority (141) of all 194 United Nations member states have, in theory, agreed to abide by the 1951 Refugee Convention. This guarantees refugees the very important right not be sent back home and put back in harm's way except under extreme circumstances.

Alongside international treaty-based structures, looser 'task-based' global groups, such as the G7 and G20, play an important influencing role too within global systems. At their annual summit meetings, the G7 group of nations have sometimes agreed to combat forms of protectionism such as import taxes and other obstacles to trade or investment. Many economists believe protectionist measures are economically harmful at both national and global scales because they deter TNCs from expanding into new markets or building cost-effective global supply chains.

This section examines how large-scale 'architectural' features of the global economy have been shaped by the work of a variety of supranational forces (IGOs, trade blocs and other global groups).

## Intergovernmental organisations and agreements

Numerous organisations exist whose membership is made up of sovereign states or their representatives. They work to facilitate intergovernmental co-operation on global trade, migration and data flows, along with important associated issues such as the human rights of refugees. In particular, the so-called Bretton Woods Institutions have played a pivotal role in world affairs by harmonising political and economic frameworks for development and investment on a planetary scale (and hence the growth of global systems).

### The work of the Bretton Woods Institutions

In the immediate aftermath of the Second World War, the groundwork was laid for three intergovernmental organisations that today exert great continuing influence on world trade and economic development. They are the World Bank, the IMF and the WTO (see Table 2.2).

Acting together as 'brokers' of globalisation through the promotion of free-trade policies and FDI, the three Bretton Woods Institutions have collectively built a global free-trade consensus. From the outset, the USA and other leading industrialised nations have strongly influenced the institutional values and priorities held by the IMF, World Bank and WTO (the former have their headquarters in Washington, DC, USA; the WTO is based in Geneva, Switzerland). This has inevitably led to accusations of undue influence by the West over the world trading system, a theme which is returned to later in this chapter.

In the late 1940s, the industrialised nations of the West sought to rebuild economic stability following the Great Depression of the 1930s and the Second World War. The 1930s were marked by mass unemployment and

Pearson Edexcel

AQA

OCR

WJEC/Eduqas

**KEY TERMS**

**Protectionism** When state governments erect barriers to foreign trade and investment such as import taxes. The aim is to protect their own industries from competition.

**Bretton Woods Institutions** The International Monetary Fund (IMF) and the World Bank. These two important organisations were founded at the Bretton Woods conference in the USA at the end of the Second World War to help rebuild and guide the world economy. The General Agreement on Tariffs and Trade (GATT) was set up soon afterwards and later became the World Trade Organization (WTO).

▲ **Figure 2.8** The headquarters of the World Bank and IMF are in Washington, DC, USA, in close proximity to the White House. One view is that the USA has disproportionate influence over the Bretton Woods Institutions

| Organisation and foundation date | Head office | Analysis of its actions affecting global flows and systems | Evaluation of its overall global importance and impacts |
|---|---|---|---|
| World Bank (1944) | Washington, DC, USA | ■ The non-profit-making World Bank provides advice, loans and grants on a global scale. It aims to reduce poverty and to promote economic development (rather than crisis support).<br><br>■ In total, the World Bank distributed US$42 billion in loans and grants in 2017. For example, help was given to the Democratic Republic of the Congo to kick-start a stalled mega-dam project, and a US$470 million loan was granted to the Philippines for a poverty-reduction programme. | ■ The World Bank has certainly succeeded in promoting trade and economic development. It has arguably helped the world avoid a return to the protectionist policies of the 1930s.<br><br>■ However, economic development can be accompanied by rising inequality also on account of neoliberal policies (see pages 145–146). The World Bank has been known to impose strict conditions on its loans and grants. Its critics describe this practice as neo-colonialism. |
| IMF (1944) | Washington, DC, USA | ■ The IMF's main function is to maintain an orderly system of lending and debt repayment between countries.<br><br>■ Under the umbrella of the UN, the IMF lends money to states in financial difficulty that have applied for assistance. The loans are supplied by other countries and must be repaid.<br><br>■ Help is provided to countries across the development spectrum when they encounter financial difficulty. For example, between 2010 and 2015, almost US$40 billion was lent to Greece to help end a period of financial crisis. | ■ IMF rules and regulations are controversial, especially the strict conditions imposed on borrowing governments. In return for help, recipients agree to run free-market economies that are open to investment by foreign TNCs. Governments may also be required to cut back on healthcare, education, sanitation or housing spending.<br><br>■ Critics say that the USA and European countries exert too much influence over IMF policies. The IMF has always had a European president and is based in the USA. |
| WTO (1995) (previously GATT, 1947) | Geneva, Switzerland | ■ The WTO took over from the General Agreement on Tariffs and Trade in 1995. Based in Switzerland, the WTO advocates trade liberalisation on a global scale – especially for manufactured goods – attempts to settle disputes and asks countries to abandon protectionist attitudes in favour of un-taxed trade.<br><br>■ For example, the WTO helped persuade China to increase its exports of 'rare earth' minerals needed for phone manufacturing by companies in other countries. | ■ Unfortunately, a round of negotiations which began in 2001 stalled for 14 years. Trying to get 162 member states to agree anything can be challenging.<br><br>■ Difficult problems for the WTO to deal with include (i) rich countries failing to agree over how far trade in agriculture should be liberalised and (ii) the fast growth of emerging economies, including China (which makes it harder to agree on fair policies for so-called 'developing' countries). |

▲ **Table 2.2** An analysis and evaluation of ways in which the Bretton Woods Institutions influence global systems

 **KEY TERM**

**Neo-colonial** A term originally used to characterise the indirect actions by which developed countries exercise a degree of control over the development of their former colonies (more recently, it has become widely used to describe some of China's overseas activities too). Neo-colonial control can be achieved through varied means, including conditions attached to aid and loans, cultural influence and military or economic support (either overt or covert) for particular political groups or movements within a developing country.

hardship for ordinary working people. The prevailing view at the Bretton Woods Conference was that protectionism led to the Great Depression. Leading industrial nations had blocked foreign imports with tariffs, in turn damaging exports for other countries. In what became a 'tit-for-tat' game, industrialised countries suffered declining economic output and accelerating political volatility during the 1930s, thereby sowing the seeds of fascism in German and Italian politics, and leading ultimately to the outbreak of war in 1939.

To avoid any return to the days of excessive trade barriers, several key principles underpinned the new consensus built at Bretton Woods.

- *From the outset, the World Bank and IMF would function as arbiters of a unified global system built around finance and trade.* Assistance from these lending organisations would be made available to help states experiencing financial difficulty or to correct economic imbalances. Over time, the remit of the World Bank and IMF has broadened to include offering long-term development assistance to low-income countries. Today, both institutions are formally part of the United Nations System (and are defined as specialised agencies of the UN) but the WTO remains independent.
- *The establishment of the GATT was intended to help remove barriers to flows of trade and investment around the world.* This goal has been pursued since then, with mixed results, by the GATT and its successor, the WTO, through a series of 'rounds' or meetings.
- *The establishment of a fixed-rate exchange system based on gold and the dollar.* The aim here was to make trade and investment easier and to help global financial flows grow over time.

Over time, the Bretton Woods Institutions have co-constructed a global legal and economic framework that enables and promotes free trade and FDI within global systems. TNCs have thrived in this environment, helped by the gradual removal of barriers to trade and investment in new markets. Privately-owned business capital has increasingly found it easier to move around the world unhindered by 'red tape'.

However, controversial borrowing rules have sometimes led to criticism of the IMF and World Bank. Since the 1970s, progressively tougher conditions have been attached to large-scale lending, including the introduction of structural adjustment programmes (SAPs) which have sometimes proved harmful to impoverished and vulnerable populations. This theme is returned to in Chapter 5.

## KEY TERMS

**Tariffs** The taxes that are paid when importing or exporting goods and services between countries.

**Structural adjustment programmes (SAPs)** Since the 1980s, the Enhanced Structural Adjustment Facility (ESAF) has provided lending but with strict conditions attached. In reality, this has meant many borrowing countries have been required to privatise public services.

# CONTEMPORARY CASE STUDY: THE NEW DEVELOPMENT BANK

▲ **Figure 2.9** The BRICS summit in 2017

Until recently, the Bretton Woods Institutions held a largely unchallenged position within global systems in terms of their financial power and influence (one possible exception to this is the Asian Development Bank, established in 1966 – but the ADB is rarely viewed as a truly 'independent voice' because the USA is one of its major donors and policy-makers). In 2014, however, the BRICS group of nations (Brazil, Russia, India, China and South Africa) announced the establishment of their New Development Bank (NDB) as an alternative to the World Bank, IMF and ADB for countries seeking financial assistance for capital-intensive development projects (see Figure 2.9).

Nobel Prize winning economist Joseph Stiglitz said the NDB represents 'a fundamental change in global economic and political power'. Certainly, it challenges the previously dominant lending model. Running in tandem with the NDB, China has also established the China Development Bank (CDB). Following the global financial crisis, the CDB loaned more than US$110 billion to developing countries in 2010, a value that exceeded World Bank lending (also see page 66).

The arrival of the NDB and CDB means that poorer nations no longer have to agree to the lending terms of the US-dominated Bretton Woods Institutions (the IMF and World Bank). This can be viewed as a step towards a more democratic world order and a challenge to the neoliberal orthodoxy of the so-called 'Washington Consensus' (meaning the shared views and principles of the Washington-based IMF and World Bank). However, the new banks have far less experience than the IMF and World Bank of managing global economic systems.

## Free trading blocs and treaties

Alongside intergovernmental organisations, numerous regional groupings of states have emerged in recent decades in the form of trade bloc agreements. The primary aim is to stimulate flows of trade, often among groups of neighbouring countries, such as the USA, Canada and Mexico (first bound together by 1994's North American Free Trade Agreement, or NAFTA), Mercosur (Argentina, Brazil, Paraguay and Uruguay) and the COMESA group of eastern and southern African nations (see page 84).

There are now over 30 major trade blocs and agreements in existence, all exhibiting varying degrees of market liberalisation and customs harmonisation. Some are composed of states at varying levels of economic development.

More ambitious trans-global partnerships have also developed or are in the process of being created. These include the following:

- *The Trans-Pacific Partnership* (TPP). This is a planned trade deal between 11 leading Pacific rim nations including Japan, Australia and Canada (the USA was originally meant to be part of this group, but the deal was abandoned by Donald Trump in one of his first acts as US president).

- *The AGOA trade pact.* The US government's African Growth and Opportunity Act offers duty-free access to lucrative US markets for qualifying sub-Saharan African countries including Lesotho and Ghana. The AGOA legislation was extended in 2015 by a further ten years to 2025.

Any decision by states to participate openly in free trade is taken knowing that the promise of easier international sales for firms sits alongside an increased risk of foreign goods flooding home markets. However, the overarching logic of the agreement dictates that all member-state companies and citizens should, on balance, become net beneficiaries of the new economic order (the economic rationale for trade bloc growth is explained in Chapter 3, pages 82–84). Thus a degree of economic sovereignty is willingly ceded by all governments. Not all citizens will support this decision, however.

Once a trade bloc has been established, free trade between neighbours or more distant allies is encouraged by the abolition or reduction of internal tariffs. Removing barriers to intra-community trade brings numerous benefits for businesses. For instance, when ten nations including Poland joined the EU in 2004, German supermarket firm Lidl gained access to 75 million potential new customers. Similarly, Sweden's IKEA has expanded its network of stores across the EU and now has over 100 mega-sized stores installed in 24 European states (while sourcing its parts and products from many of the same countries).

In reality, however, the EU and other trade blocs are far from being completely 'borderless'. A multitude of legal and economic obstacles to *entirely* free trade and investment remain. For example, the French and Italian governments monitor, and potentially halt, unwanted foreign corporate takeovers in sectors deemed 'strategically important', such as energy, defence, telecoms and food.

Politicians must constantly reappraise the real or perceived costs and benefits of trade bloc membership. The sheer complexity of cross-border investment and flows complicates this task greatly. As a result, perspectives usually vary among politicians and the citizens they represent about the wisdom of trade agreements. Views are coloured by personal experiences too: some US citizens blame NAFTA for their own unemployment, and Donald Trump's 2016 presidential campaign championed this view. His call for a border wall with Mexico was well received by US citizens who want stronger barriers against illegal immigration and foreign imports.

 **KEY TERM**

**Sovereignty** The ability of a place and its people to self-govern without any outside interference.

## Other influential global groups

Alongside IGOs and trade blocs, other kinds of global group also play a significant supranational role in world affairs, particularly in relation to trade and investment. Table 2.3 analyses and evaluates the influence of three of these powerful groups over global systems and global flows.

| Groups | Analysis | Evaluation |
|---|---|---|
| G7/8 and G20 | ■ The G7 'Group of Seven' nations comprises the USA, Japan, UK, Germany, Italy, France and Canada (conferences held with Russia prior to 2014 were called G8 meetings). Since 1975, the world's largest economies have met periodically as a kind of 'task force' to co-ordinate their response to common economic challenges.<br><br>■ In 2011, the G8 acted to stabilise Japan's economy after the devastating tsunami. In 2016, the G7 met to discuss policies capable of stimulating growth in response to the global economic drag caused by China's slowdown. | ■ The G7 is steadily becoming less important as a forum for international decision making. This is because several leading economies, including China, India, Brazil and Indonesia, are not G7 members. A larger group called the G20 has therefore been established which includes these leading emerging economies in addition to the G7 members and Russia.<br><br>■ The larger size of the G20, and the differing views of its members, sometimes weakens its ability to agree and act on issues. Recently, tensions between Russia and EU states have become more evident. |
| Organisation for Economic Co-operation and Development (OECD) | ■ The OECD is a forum of 36 high-income and middle-income countries. Its mission is 'to promote policies that will improve the economic and social wellbeing of people around the world'.<br><br>■ Member states have signed formal agreements on protecting the environment.<br><br>■ They have agreed to work together to tackle the challenge of ageing populations. | ■ The OECD has made good progress towards clamping down on tax evasion by TNCs. Rules to stop companies using complex tax arrangements to avoid paying corporate tax have been agreed by 31 members. It will be harder for firms to hide money in tax havens in the future.<br><br>■ However, OECD economists completely failed to predict the slowdown in the world economy which began in 2008 (see page 91). This was a huge oversight. |
| Organization of the Petroleum Exporting Countries (OPEC) | ■ OPEC is a wealthy and important global cartel of oil-producing countries which includes Saudi Arabia and Qatar. As demand for oil has grown, OPEC nations have gained enormous wealth.<br><br>■ Global dependency on oil ensures OPEC countries are key political players, with real influence on the world stage. | ■ Several OPEC countries have suffered the destabilising effects of civil war, insurgency or international conflict, including Kuwait, Iraq and Nigeria. There is one view that oil can actually hinder rather than help a country's development ('the oil curse').<br><br>■ The collapse in world oil prices in 2015 left several OPEC members, including Nigeria and Venezuela, in financial difficulty. |

▲ **Table 2.3** Powerful and influential global groups of countries

### Global interactions with TNCs

IGOs and other global groups act in ways that sometimes stimulate global trade and investment flows; they may also seek to control, change and regulate the ways TNCs operate, resulting in an at times complex interrelationship between political and economic players. Both the OECD and G20 groups want stricter controls imposed on the way TNCs use transfer pricing mechanisms and other tax avoidance strategies (see page 45).

- Since 2009, the 36-strong OECD has closely monitored offshore tax havens that are believed to enable US$7 trillion tax evasion on global TNC profits each year.
- In 2015, a G20 project culminated in 60 governments agreeing to get tougher on tax evasion and 'profit shifting' by TNCs. Although the Cayman Islands and Bermuda remain legitimate places to register a business, many TNCs are beginning to question the wisdom of operating there because of increased OECD scrutiny of their financial networks. Firms are more aware of the brand risk associated with profit shifting via Ireland, the Netherlands or other low-tax destinations (see Figure 2.4, page 45).

## International migration agreements

Within the EU, free movement of labour is permitted. Southern England, northern France, Belgium and much of western Germany are important host regions for much of the migration which has occurred. This area includes the world cities of London, Paris, Brussels and Berlin. Labour migration flows from source regions in eastern and southern Europe are overwhelmingly directed towards these places. Most national border controls within the EU were removed in 1995 when the Schengen Agreement was implemented. This enables easier movement of people and goods within the EU, and means that passports usually need not be shown at borders between EU countries.

Other world regions have begun to adopt free-movement rules too.

- South American countries have also taken steps towards this goal. Between 2004 and 2016, around 2 million South Americans obtained a temporary residence permit in one of nine countries implementing the agreement. After the signing of the Mercosur Residence Agreement, nationals of Argentina, Bolivia, Brazil, Chile, Colombia, Ecuador, Paraguay, Peru and Uruguay have the right to apply for temporary residency in another member state. After two initial years of temporary residency, it is possible to convert the status to permanent residency.
- The African Union has said it wants to break down borders through closer integration. In 2016, the African Union (which has 55 member states) began issuing e-passports that permit recipients to enjoy visa-free travel between member states.

### Rules governing refugees

Refugees are people who have been forced to leave their country. They are defined and protected under international law, and must not be expelled or returned to situations where their life and freedom are at risk. In addition to refugees, many people worldwide have become internally displaced persons (IDPs) after fleeing their homes. In 2016, the conflict in Syria that started in

2011 had generated 5 million refugees and 6 million IDPs; half of those affected were children.

According to UN data:

- In 2014, more people were forced to migrate than in any other year since the Second World War. Fourteen million people were driven from their homes by natural disasters and conflict. On average, 24 people were forced to flee for their lives each minute, four times more than a decade earlier.
- The global total of displaced people now exceeds 65 million. Of these, around 45 million are internally displaced and 20 million are refugees.
- Recent forced movements of people have been caused by wars in Syria, South Sudan, Yemen, Burundi, Ukraine and Central African Republic. Thousands more have fled violence in Central America. Figure 2.10 shows the list of countries that have generated most refugees on a rolling annual basis since the 1970s.

Some of these movements are caused by pressures created by global systems (see pages 153 and 158). In theory, most countries are obliged to take in refugees, irrespective of whatever economic migration rules exist. This is because they have signed the Universal Declaration of Human Rights (UDHR), which guarantees all genuine refugees the right to seek and enjoy asylum from persecution (see Table 2.4).

| The Refugee Convention (1951) and Convention relating to the Status of Stateless Persons (1954) | ▪ The 1951 Refugee Convention is the key legal document that forms the basis of all UN work in support of refugees. Ratified by 145 states, it defines the term 'refugee' and outlines the rights of refugees, as well as the legal obligations of states to protect them. The core principle is 'non-refoulement'. This means that refugees should not be returned to a country where they face serious threats to their life or freedom. This is now a core rule of international law.<br><br>▪ The 1954 Convention relating to the Status of Stateless Persons was designed to ensure that stateless people enjoy a minimum set of human rights. It established human rights and minimum standards of treatment for stateless people, including the right to education, employment and housing. |
|---|---|
| The United Nations High Commissioner for Refugees (UNHCR) | ▪ The UNHCR serves as the 'guardian' of the 1951 Refugee Convention and other associated international laws and agreements. It has a mandate to protect refugees, stateless people and people displaced internally. On a daily basis it helps millions of people worldwide at a cost of around US$5 billion annually. The UNHCR works often with the UN's World Health Organization (WHO) to provide camps, shelter, food and medicine to people who have fled conflict.<br><br>▪ The UNHCR also monitors compliance with the international refugee system. |

▲ **Table 2.4** How the United Nations offers protection to refugees

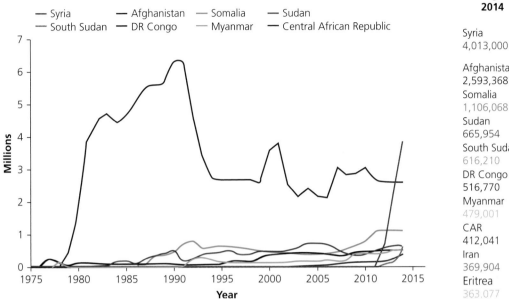

**Highest number 2014**

Syria
4,013,000

Afghanistan
2,593,368

Somalia
1,106,068

Sudan
665,954

South Sudan
616,210

DR Congo
516,770

Myanmar
479,001

CAR
412,041

Iran
369,904

Eritrea
363,077

▲ **Figure 2.10** The main source countries for refugees, 1975–2014

## Multi-scalar governance

Given the complexity of global systems, how much influence can one single government or intergovernmental organisation hope to have over patterns of trade, migration or investment? The answer is often: very little. As a result, geographical research often focuses on how *interactions* between different national governments and IGOs affect (i) how global systems evolve over time, and (ii) outcomes for individual nations.

As an illustration of this, Figure 2.11 shows a timeline of Indonesia's integration into global systems.

- Today, it is the world's seventh-largest economy, measured by purchasing power parity (PPP). But just 50 years ago, Indonesia's government was strongly opposed to the Bretton Woods Institutions and looked set to follow a path similar to that of North Korea today.
- The opening up of Indonesia's economy to global flows in the late 1960s occurred only after the country's new rulers chose to work alongside the IMF and Anglo-American TNCs to create a new 'investment regime'.
- This collaboration between national (state) and international players (the IMF, along with private-sector TNCs) paved the way for Indonesia's entry into global systems (and the costs and benefits this subsequently created for its people).

 **KEY TERM**

**Purchasing power parity (PPP)** A measure of average wealth that takes into account the cost of a typical 'basket of goods' in a country. In low-income countries, goods often cost less, meaning that wages go further than might be expected in a high-income country.

| Globalisation phases | Factors |
|---|---|
| **Phase 1: Colonised by the Dutch**<br>1670–1945   For nearly 300 years, Indonesia experienced an early form of globalisation as an exploited and dependent European colony. It was also invaded by the Japanese during the Second World War. | • **Raw materials** drew the Dutch to Indonesia. Flows of valuable tin, copper, timber, rubber and gold made Indonesia's islands an important prize. |

| Globalisation phases | Factors |
|---|---|
| **Phase 2: Independence and anti-Western stance**<br>1946–64   After four years of guerilla warfare, new leader General Sukarno took power, with a strong anti-Western stance. The message 'Go to hell with your aid' was sent to the USA and the IMF. Sukarno courted the communist Soviet Union. | • **De-colonisation** left Indonesia viewing the free-market Western world with suspicion.<br>• **The Cold War** (when the USA and the Soviet Union were opposing world superpowers) was a major influence on the globalisation of Indonesia. |

| Globalisation phases | Factors |
|---|---|
| **Phase 3: Regime change and pro-Western stance**<br>1965–67   Backed by Washington, General Suharto seized power from General Sukarno during a period that saw hundreds of thousands of suspected Indonesian communists murdered.<br>1968   Suharto's new regime opened up Indonesia's economy. American and European TNCs met with the Suharto government in Switzerland and designed new attractive economic policies for free-market investors coming to Indonesia. | • **IMF infrastructure loans** Western TNC branch plants arrived in Indonesia as soon as its roads, power supplies and ports began to modernise.<br>• **Legal changes** The legal framework for the export processing zone in Jakarta gave TNCs exactly what they needed – a low-tax haven for sweatshop manufacturing, making the most of low labour costs. |

| Globalisation phases | Factors |
|---|---|
| **Phase 4: Economic collapse and recovery**<br>2000+   Poor publicity resulted in many foreign investors like Gap Inc. improving conditions for workers. After nearly 50 years as a major global player, Indonesia is the world's 7th largest economy by one measure (2018 data). However, progress still needs to be made for Indonesia's poor (the country is ranked just 115th when GDP per capita is measured). 40% of the population still lives below or just above the poverty line. | • **Anti-globalisation** protests have put the spotlight on Indonesia's sweatshops and the tough line taken by the Suharto regime against trade union campaigners, many of whom were imprisoned. Half the population live on less than US$2 a day.<br>• **G20 membership** has made Indonesia a major world political player, adding another dimension of globalisation to the nation's profile. |

▲ **Figure 2.11** Indonesia's global systems timeline: can you see how events have been shaped by 'multi-scalar' interactions between Indonesia's government, IGOs and powerful TNCs?

◄ **Figure 2.12** Indonesia today is the world's seventh-largest economy. However, the country's integration into global systems almost did not happen

# 3 The influence of superpower states

▶ *How have unequal power relations allowed some states to drive global systems to their own advantage?*

Global systems, as this chapter has already shown, are partly shaped by national government decision-making within frameworks fashioned by IGOs. Additionally, a small handful of extremely powerful countries exert undue influence over other nations' governments and IGOs alike. These are the superpower states. The USA, China and arguably Russia are true global superpowers with military capabilities to match. A case can also be made that Japan and some European states, including the UK and France, are superpowers, or have been in the recent past.

Powerful countries use their influence to establish global economic and political frameworks that suit their own needs well. The Bretton Woods Institutions, for example, were established in a large part through the efforts of the USA; they have subsequently promoted free trade in ways that are extremely beneficial for US TNCs. Superpower states will also use diverse strategies and tactics to protect a *status quo* they have helped establish; hence, geographers will sometimes refer to the way global systems are 'produced and reproduced' by powerful countries. The meaning here is that superpower states protect and maintain the international institutions and laws they created; thereby ensuring the survival of global systems which, in their current form, deliver ongoing benefits to the superpowers themselves.

## Using soft and hard power to reproduce systems

The term 'global superpower' was used originally to describe the ability of the USA, USSR and the British Empire to project power and influence anywhere on Earth to become a dominant worldwide force.

- The British Empire was a colonial power, alongside France, Spain, Portugal and other European states. Between approximately 1500 and 1900, these leading powers built global empires. One result was the diffusion of European languages, religions, laws, customs, arts and sports on a global scale.
- In contrast to the direct rule of the British in the 1800s, the USA has dominated world affairs since 1945 mainly by using indirect forms of influence or neo-colonial strategies. These include the US government's provision of international aid and the cultural influence of American media (including Hollywood and Facebook). Alongside such soft power strategies, the US government has routinely made use of hard power.

 **KEY TERMS**

**Soft power** The political scientist Joseph Nye coined the term 'soft power' to refer to the art of persuasion. Some countries are able to make others follow their lead by making their policies attractive and appealing. A country's culture (arts, music, cinema) may be viewed favourably by people in other countries. Nye noted that one can influence the behaviour of others in three main ways: threats of coercion ('sticks'); inducements and payments ('carrots'); and attraction (the most subtle form of influence). The last two of these are soft power.

**Hard power** This means getting your own way by using force. Invasions, war and conflict are very blunt instruments. Economic influence can be used as a form of hard power: sanctions and trade barriers may cause great harm to other states.

This means the geopolitical use of military force (or the threat of its use) and the economic influence achieved through forceful trade policies including economic sanctions or the introduction of import tariffs. The term 'smart power' is used to describe the skilful combined application of both soft and hard power in international relations (see Figure 2.13).

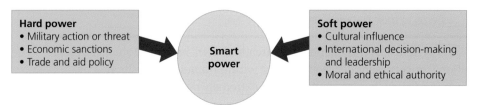

| Hard power | Soft power |
|---|---|
| • Military action or threat | • Cultural influence |
| • Economic sanctions | • International decision-making and leadership |
| • Trade and aid policy | • Moral and ethical authority |

▲ **Figure 2.13** The ingredients of 'smart power'

Other than the USA, what other states can claim to be true global superpowers?

- China became the world's largest economy in 2014 and exerts great influence over the global economic system through its sheer size. Other emerging economies including India and Indonesia play an increasingly important global role.
- Although no single European country can equal the influence of the USA, several have remained significant global players in the post-colonial world (notably the G7 nations of Germany, France, Italy and the UK). Another view is that European states can only rival the USA's global superpower status when they work together as members of the EU.

## Sovereign wealth funds

Alongside the global superpowers, we can identify a second tier of extremely influential countries who often manage to 'punch above their weight' in relation to, say, their small population size or land area. Qatar is an excellent example: this tiny Gulf state gained international influence through use of its enormous oil- and gas-derived wealth to fund substantial infrastructure projects in the UK, Tunisia, Egypt, Turkey and other countries.

These smaller but nonetheless influential countries often make their mark in the world on account of large sovereign wealth funds (SWFs) which generate vast global capital flows (see Figure 2.15, page 64). Only a minority of countries operate SWFs. These are mostly countries with:

- oil and gas revenues (such as Norway and Qatar)
- mineral resources (Chile's copper wealth and Botswana's diamonds)
- a balance of payments surplus (when a country's exports exceed the value of its imports, e.g. China).

Two different spending models are used. Direct state purchasing sometimes takes place (Kuwait's US$500 billion fund, dating back to 1953, works this way). Alternatively, some states have purpose-built investment banks which manage their purchasing (Singapore has two, called Temasek and GIC).

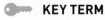

**KEY TERM**

**Sovereign wealth funds (SWFs)** These are the global-scale 'piggy banks' which some states rely on to build global influence and diversify their income sources.

Most rich and powerful developed countries, such as Japan, the UK and USA, do not operate SWFs. Interestingly, however, several US states, including Texas and Alaska, have their own wealth funds.

## The SWF shopping list

The UK is the most popular world destination for SWF investment (see Figure 2.14). Foreign states, notably China, already own major stakes in Britain's railway lines, airports, water companies, sewers, central business districts and half of the House of Fraser department store chain. Chinese SWFs are likely to invest heavily in the planned £55 billion HS2 railway line between London and Manchester. These high-cost purchases have strong potential to turn a profit in the future.

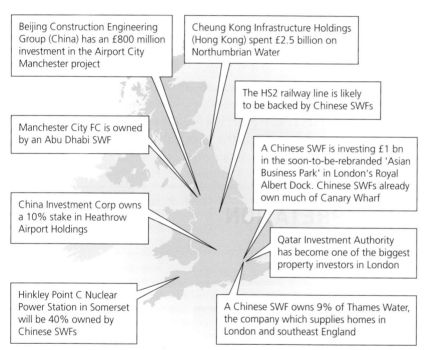

Beijing Construction Engineering Group (China) has an £800 million investment in the Airport City Manchester project

Cheung Kong Infrastructure Holdings (Hong Kong) spent £2.5 billion on Northumbrian Water

The HS2 railway line is likely to be backed by Chinese SWFs

Manchester City FC is owned by an Abu Dhabi SWF

A Chinese SWF is investing £1 bn in the soon-to-be-rebranded 'Asian Business Park' in London's Royal Albert Dock. Chinese SWFs already own much of Canary Wharf

China Investment Corp owns a 10% stake in Heathrow Airport Holdings

Qatar Investment Authority has become one of the biggest property investors in London

Hinkley Point C Nuclear Power Station in Somerset will be 40% owned by Chinese SWFs

A Chinese SWF owns 9% of Thames Water, the company which supplies homes in London and southeast England

◀ **Figure 2.14** UK assets acquired by overseas SWFs. This shows that not only foreign TNCs invest in the UK; increasingly, foreign governments invest there too, bringing greater complexity to global systems and capital flows

After the global financial crisis (GFC) of 2008–09 (see page 92), the UK government actively encouraged investment in the UK by overseas SWFs. This was because of a shortfall of money for new projects (without raising taxes or the UK's national debt). Yet the government's very own National Infrastructure Plan identified the need for half a trillion pounds of spending on transport, energy and cities before 2020. In order to get these big projects off the drawing board, the government was obliged to look to overseas investors for support.

- Speaking in 2014, UK Prime Minister David Cameron said 'I'm not embarrassed that [China] own ten per cent of our biggest water company [Thames Water] or a big chunk of Heathrow airport. I'm proud. Tell the other Chinese investors – come to London; spend your money.' By 2025, China will own an estimated £100 billion of UK energy, property and transport investments.

- Democracy and rule of law make the UK a low-risk investment site for Russian, Singaporean and Middle Eastern SWFs. All have spent heavily on UK assets. Perhaps the best-known example is Sheikh Mansour bin Zayed Al Nahyan of Abu Dhabi's ownership of Manchester City football club (more than £1 billion has been invested, including the cost of a state-of-the-art training academy).
- Not everyone agrees SWF investments are a good thing, however. Because ownership passes to foreign governments, rather than foreign companies, critics say SWF purchases represent a 'loss of sovereignty'. In effect, the UK government is ceding power over national assets to foreign governments (some of whom do not operate democracies).

### Investing around the world

The UK is not the only place where SWFs invest, of course.

- Singapore owns almost half of New Zealand's Viaduct Quarter (a major residential and commercial development in Auckland).
- Angola is investing its US$5 billion oil-derived assets on infrastructure and hotels in neighbour states.
- Norway, the world's largest SWF, is a major Facebook shareholder.
- SWFs play an important role in the so-called global land-grab phenomenon (see page 158).

## ANALYSIS AND INTERPRETATION

Study Figure 2.15, which shows the relative size of different oil and non-oil sovereign wealth funds.

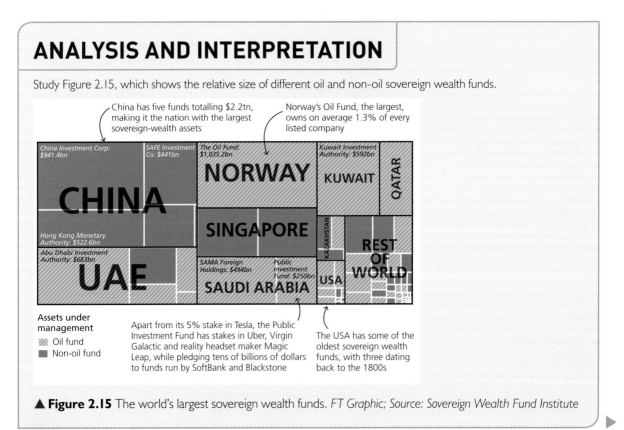

▲ **Figure 2.15** The world's largest sovereign wealth funds. *FT Graphic; Source: Sovereign Wealth Fund Institute*

(a)    Estimate the total value of the UAE's sovereign wealth funds.

**GUIDANCE**

In order to answer this accurately, we need to look carefully at how the data have been presented. Thin white lines separate the value of different funds belonging to the same country: Singapore has two funds, for instance. The UAE (the focus of this question) has several funds, one of which is valued at US$683 billion. Measure the width of that fund (as represented in Figure 2.15) and also the width of the remainder of the box (a shape made up of the remaining four funds). You can use these two widths to calculate a ratio which can in turn be used to calculate the total value of the UAE's sovereign wealth funds.

(b)    Suggest reasons why only a small number of countries control most of the world's largest sovereign wealth funds.

**GUIDANCE**

This question can be answered in two different ways. First, there will be specific reasons why the countries shown have unusually large financial assets: what are the likely sources of wealth for the non-oil funds? Apply your knowledge and understanding of different kinds of global trade to suggest reasons why China has so much money to invest, for example. Second, there are political reasons why some governments choose to operate sovereign wealth funds, whereas others do not. This is often related to the kind of political system found in different countries. For instance, in communist China the state controls many of the country's industries. As a result, large financial assets built from the profits of global trade are available for the Chinese government to use as it sees fit.

(c)    Explain ways in which the sovereign wealth funds shown in Figure 2.15 play a role in global systems.

**GUIDANCE**

One way to approach this question is to think about the different global flows that operate within global systems, including flows of money, merchandise, services, people and ideas. Using examples provided in this chapter, you can offer a sequential explanation of the role played by sovereign wealth funds in relation to each of these flows. For example, large investments by the UAE and China around the world may require people from those countries to work overseas managing projects: flows of people therefore accompany the flows of money.

# CONTEMPORARY CASE STUDY: QATAR

The tiny Middle Eastern kingdom of Qatar has the highest per capita GDP in the world, in excess of US$100,000.

2.5 million people live in Qatar but only 300,000 of these are Qatari citizens. The remainder are a mixture of low- and high-skilled migrants. There is a huge demand for construction site workers in Doha, the capital city (see Figure 2.16). Large flows of professional (or 'elite') migrants are directed towards Qatar too

because it has become a global hub for investment, culture and media industries.

The country's wealth and global influence, like that of neighbouring Saudi Arabia, stems originally from fossil-fuel wealth: Qatar has 14 per cent of all known gas reserves. The Qatari government has subsequently reinvested its petrodollar wealth in ways that have diversified the national economy and built global influence too.

- The capital city of Doha has become a globally renowned place where international conferences and sporting events are held, served by Qatar Airways and Doha International airport. Important UN and WTO meetings have taken place in Doha, including the 2012 United Nations Climate Change Conference with about 17,000 participants. Before this, Qatar hosted the WTO conference known as the Doha Round of talks in 2001. The city is set to host the 2022 Football World Cup.

- Long-standing global Anglo-American media dominance was broken by the arrival of Qatar's Al Jazeera media network in 1996. Almost from the outset, it has rivalled the BBC and CNN for influence in many parts of the world. This universally recognisable brand is an important source of soft power for Qatar. You can tune in and watch it on Freeview in the UK.

- The Qatar Investment Authority (QIA), one of the world's largest SWFs, owns a number of 'trophy assets' in the UK, including the Shard building, Harrods department store and a major stake in the London Stock Exchange. The QIA poured almost £3 billion into UK real estate and infrastructure in 2018.

- Qatar hosts the biggest US military base in the Middle East, making it an important ally of the world's greatest superpower.

However, many people regard Qatar as a regional power, rather than a true global superpower in its own right. Another view is that there are limits to Qatar's soft power because it is an autocratic and authoritarian state where people's human rights are not always protected. Also, since 2017, Qatar has been involved in a serious diplomatic dispute with Saudi Arabia, the United Arab Emirates, Egypt and Bahrain. These countries have accused Qatar of financing terrorism and supporting Islamist groups. This has adversely affected the country's ability to project its influence globally.

▲ **Figure 2.16** Qatar's capital city, Doha

# CONTEMPORARY CASE STUDY: THE BELT AND ROAD INITIATIVE

Chinese sovereign wealth funds have bankrolled new infrastructure projects in 78 countries between 2013 and 2018. This ambitious global development programme is called the Belt and Road initiative. China's leader, Xi Jinping, has described it as the start of 'a new golden age of globalisation'. Critics of China see it as a cynical attempt to gain global political power and influence in return for high-risk loans. Conversely, Belt and Road Initiative supporters say that increased south–south lending is good news because it indicates the emergence of a new multi-polar world (meaning that political and economic power are no longer concentrated solely in the hands of the USA and its allies).

According to the World Bank, some Belt and Road Initiative countries are politically and/or economically high-risk states for investment (see Figure 2.17). One

view is that China's rulers are content to allow foreign governments in Africa and South America to borrow more money than is sensible. This is because the Chinese government believes it will gain political leverage over any countries which default on their loans. If debtors cannot repay the money lent, this lets China gain control of whichever strategic assets the loans have paid for, such as airports, power stations or other vital infrastructure.

- In 2018, the IMF warned that China's loans to Pacific nations could trigger a new debt crisis after Beijing committed US$6 billion to projects since 2011.

- Pakistan in particular has borrowed so heavily from China that in 2018 it was almost unable to repay debt interest (let alone the actual value of the debt itself).

- Other countries in difficulty included Sri Lanka, Montenegro (after borrowing almost US$1 billion to fund a motorway project) and Laos (which borrowed US$6 billion – almost half of its annual GDP – for a railway project).

- In 2018, Malaysian Prime Minister Mahathir Mohamad threatened to cancel China-funded projects worth US$20 billion, and warned: 'There is a new version of colonialism happening.'

Alongside the Belt and Road Initiative, additional large capital outflows from China in recent years include the following.

- *New funds for African energy.* In 2017, the China Development Bank and the Export – Import Bank of China (the world's two largest development banks) lent six African countries (Angola, Nigeria, Zambia, Uganda, South Africa and Sudan) around US$7 billion for power projects.

- *New European investments.* China is growing the value of its overseas acquisitions in Europe, including almost 300 significant investments made in Germany during 2016.

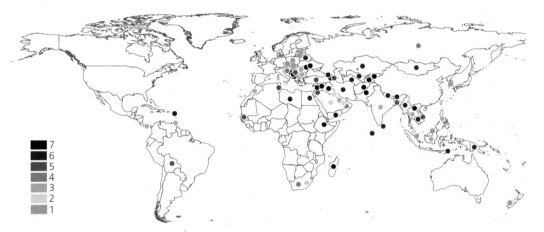

7
6
5
4
3
2
1

▲ **Figure 2.17** Belt and Road Initiative countries are shown here according to how poorly they are ranked by the OECD in terms of investment risk (due to economic mismanagement or political instability). Critics of China say the country has cynically lent large sums of money for 'white elephant' projects in countries that are likely to struggle to repay their debts, thereby leaving them indebted to China. Like the Bretton Woods Institutions (who critics say are guilty of imposing neoliberalism on developing countries), is China too exploiting global financial systems for its own advantage?. *FT Graphic Jane Pong, James Kynge; Sources: OECD, FT research*

#  Evaluating the issue

▶ *To what extent does globalisation involve the imposition of Western ideas on the rest of the world?*

## Identifying possible contexts and criteria for the evaluation

The final section of this chapter is concerned with power over *ideas* within global systems. To what extent do powerful 'Western' countries have disproportionate influence when it comes to global flows of ideas? Is a global culture emerging based mainly on Western ideas? Or are outcomes actually more complex?

We need to think critically about what is possibly meant by the statement: 'globalisation involves the imposition of Western ideas on the rest of the world'. Three underlying assumptions need to be addressed.

1 Globalisation, as Chapter 1 showed, is a complex multi-strand concept involving multiple processes of change; it has economic, social, cultural and political dimensions. Not all globalising processes necessarily lead to the transmission of ideas. For example, raw material flows – including primary commodities like oil and timber – may have little or nothing to do with the diffusion of ideas and values. In contrast, flows of people (migration) and data (including streamed media and social networking) are more likely to be 'carriers' of cultural traits and norms (see Figure 2.18).

2 What is meant by 'Western ideas' is not entirely clear. 'Western' is sometimes used as a synonym for the developed countries of Europe and North America, along with Australia and New Zealand (Japan, although a developed country, is not 'Western'). Whether or not all eastern European countries are fully 'Western'

in character is debatable. Moreover, as this chapter has shown, the USA has disproportionate power and influence far beyond that of many smaller European states; one view is this has led to the so-called Americanisation of world culture. But how and why might this differ from Westernisation?

3 Finally, the word 'imposition' requires careful handling. The phrase 'cultural imperialism' is used sometimes to describe the imposition in past times of cultural change on countries by colonial rulers backed with military might. We can think back to when European countries forced their languages and Christian beliefs onto people in Africa, Asia and South America. Today, cultural diffusion occurs in more subtle and less obviously coercive ways. The USA, UK, France and Russia exert great cultural influence globally via their media institutions, including Hollywood, Netflix, the BBC and RT (Russia Today). European languages remain widely spoken; English in particular is viewed as an important business language. But is it correct to view the persisting influence of Western culture and languages as a form of 'imposition'?

| **Language:** Some countries have a single national language with local dialects, or several languages belonging to different indigenous ethnic groups | **Food:** National dishes and diet traditionally reflect the crops, herbs and animal species that are available locally | **Clothing:** National and local traditions may reflect traditional adaptations to the climate (such as wearing fur in polar climates) or religious teachings | **Religion:** There are several main world religions, each with its own local variants; religion is an important cultural trait that also informs food and clothing, and may be highly resistant to change | **Traditions:** Everyday behaviour and 'manners' are transmitted from generation to generation, from parents to their children, such as saying 'thank you' or shaking hands |

**Cultural traits**

▲ **Figure 2.18** Cultural traits: which of these are more or less likely to change over time on account of global flows of goods, people and ideas?

# Evaluating evidence for the imposition of Western ideas on the world

Cultural changes linked with the spread of ideas and information are a vital aspect of globalisation. One view is that the growth of a global culture which is essentially 'Western' in character has led to reduced cultural diversity worldwide. One particularly striking manifestation of Western influence is the vast numbers of people around the world speaking Spanish, English, Portuguese and Russian. At the same time, one-in-four of the world's 7000 minor languages are threatened with extinction (see Figure 2.19); half have fewer than 10,000 speakers remaining and are spoken by just 0.1 per cent of the world's population.

Linguistic diversity worldwide has declined by about 30 per cent since 1970. Papua New Guinea originally had over 1000 indigenous languages. But as globalisation accelerates, the physical, technological and economic barriers that once allowed so many isolated languages to develop have been removed and many languages are now lost.

English in particular has thrived globally in a reduced form called 'Globish'. In 1995, Jean-Paul Nerrière first used the term to describe a stripped-down vocabulary consisting of just 1500 English words but spoken by up to 4 billion people. The Globish 'micro-language' is distinct from more complex variants of true English spoken as an official language in the USA, Australia, Canada and elsewhere. It serves a purely utilitarian purpose by providing global citizens with the means they need to exchange vital information with one another, such as travel directions or terms of business. Global citizens are people who are routinely involved in global interactions, including:

- tourists, international migrants and the international business community
- residents in global hubs (megacities such as Los Angeles or São Paulo), where many different ethnic and migrant groups require a common language for communication

▲ **Figure 2.19** The distribution of threatened and disappearing languages. Western languages like English and Spanish may play a role in their decline

- social network users, such as Facebook members, who correspond online with people of many nationalities.

Globish has a long history of being adopted by (i) international migrants arriving in English-speaking countries such as the USA and (ii) the citizens of more than 60 ex-British colonies. Since the 1990s, however, Globish has diffused into countries that traditionally lack a strong affinity with British or American culture, such as Japan, China or Brazil. This is because English:

- has dominated internet communication from its outset
- has enjoyed a supranational rise as the global language of business (commerce, technology and education) and media (music and film), in part due to the English-speaking USA's superpower status.

## The cultural influence of Western TNCs and media

Many people's idea of a typical 'global citizen' is someone who wears jeans, listens to rock or rap music, uses social media on an iPhone or Galaxy and enjoys buying branded Nike and Adidas clothing. Based on this evidence, Western TNCs and media corporations – especially those originating in the USA – do appear to exert enormous cultural influence.

- By expanding into new markets, Western TNCs have helped spread European and North American modes of food, music, clothing and other commodities. Nike, Apple and Lego have rolled out uniform products worldwide, thereby bringing greater cultural homogeneity to different places.
- The USA's media giants, especially Disney (which also owns the Marvel franchise) have exported stories about superheroes and princesses to every continent, along with films about Christmas (which was originally a Western Christian festival).

- The BBC helps the UK maintain its long-standing high level of global cultural influence and soft power.

Critics of the West might argue that cultural influence is being achieved through subtle forms of coercion. According to this view, European and North American global brands are to blame for a new, neo-colonial phase of Western cultural imperialism; in other words, Western culture is spread worldwide via a global media over which the USA in particular has a disproportionate influence. In movies and online, children in Asia, Africa and Latin America are exposed to Western traditions including Christmas (see Figure 2.20), Halloween or Valentine's Day (see page 34).

Many people view the West's cultural influence as a good thing, however. Companies like Disney, Netflix, Amazon and the BBC may sometimes be helping to spread progressive social ideas with their media output. Strong female role models and positive LGBT portrayals often feature in contemporary programming. Some people argue that these are important messages for parts of the world where human rights are not respected for women and minority groups, for example in Chibok in northeast Nigeria, where schoolgirls were sold into slavery by the Boko Haram militia group (see also page 180).

▲ **Figure 2.20** Christmas is a Western tradition that has also been embraced in many non-Christian Asian countries such as China

# Evaluating alternative views about globalisation, culture and ideas

Several arguments can be made that *counter* the claim that globalisation involves the imposition of Western ideas on the rest of the world.

First, the word 'imposition' implies a lack of choice. Yet, after gaining their independence in the 1950s and 1960s, many ex-British colonies *chose* to remain in the British Commonwealth and to retain the Anglicised names given to those countries under British rule. To this day, the Union Jack appears on the flags of several territories around the world, including Fiji and Bermuda. No overt force is involved in the exercise of Western soft power (it seems unlikely that anyone has ever been forced at gunpoint to eat a Big Mac). Moreover, the actions of TNCs cannot be classed as cultural imperialism *because most businesses operate independently of government*. US-based TNCs are motivated by profit, not politics. The guiding principle of these firms is to build their market share, and any cultural changes they bring to places happen only with their consumers' consent. Words like 'imposition' and 'imperialism' are misnomers for more subtle and consensual processes of cultural change.

Second, any spread of culture within global systems rarely occurs in a 'linear' way – ideas are not simply transmitted and received like electricity passing through a circuit board. It is more the case that flows of ideas *meet and interact with* the cultures that already exist in different places. As a result, unexpected changes happen, resulting in new cultural hybridity. Globish, for instance, has so many different local variant forms that it perhaps enhances – rather than erodes – worldwide linguistic diversity. Words, syntax and grammar vary from country to country because of the way English has blended with different native languages – 'Singlish' is the Singaporean variant of Globish, for instance, while 'Hinglish' is the version used by Hindi speakers. Also, given its highly limited vocabulary, Globish (in all its variant forms) is not *replacing* other languages; instead, people adopt it *in addition to* their first language. We should be therefore cautious about viewing Globish as an enforced change which somehow robs other places of their own identity.

Thirdly, the global spread of branded commodities also often involves a process of hybridisation called glocalisation (see also page 25) which calls into question reductive ideas about the 'imposition' of Western culture. Glocalisation involves TNCs adapting their products for different markets to take account of local variations in tastes, customs and laws. This strategy developed originally from the need of some TNCs to source parts and ingredients locally when establishing branch plants overseas. SABMiller, a major TNC, uses cassava to brew beer in Africa, for instance; this cuts the cost of importing barley (which is used in other world regions). Glocalisation makes business sense too because of geographical variations in the following:

- *People's tastes*. European TNC GlaxoSmithKline re-branded its energy drink Lucozade for the Chinese market with a more intense flavour. Working in partnership with Uni-President China Holdings, the product's new local name translates as 'excellent suitable glucose'.
- *Religion and culture*. Domino's Pizza only offers vegetarian food in India's Hindu neighbourhoods; MTV avoids showing overtly sexual music videos on its Middle Eastern channel.
- *Laws*. The driving seat should be positioned differently for cars sold in the US and UK markets.
- *Local interest*. Reality TV shows gain larger audiences if they are re-filmed using local people in different countries.

Driving this attention to detail is the sheer size of emerging markets. Strengthening Latin American, Asian, African and Middle Eastern economies are home to growing numbers of cash-rich young people in places previously sidelined by TNCs. Leading global brands want to maximise their sales chances. Glocalisation has therefore become a vitally important economic, political and cultural strategy that informs and defines certain kinds of companies' actions in the global market place (see Table 2.5). Increasingly, TNCs are listening to what customers in different markets tell them they want. They are not seeking to impose a uniform product or service.

## Evaluating the view that Western ideas are changed and challenged by other cultures

Increasingly, non-Western ideas are influencing how global culture evolves. According to this view, the idea that globalisation and Western influence are one and the same thing is actually very naive. Earlier in this chapter we explored the growing influence of new global and regional powers including China, India, Japan, Brazil and Qatar (see page 65). Times have changed and there are now plenty of non-Western ingredients in the melting pot of global culture. Important non-Western influences on global culture include

| McDonald's Corporation | MTV (Music Television) | The Walt Disney Company |
|---|---|---|
| By 2015, McDonald's had established 35,000 restaurants in 119 countries. In India, the challenge for McDonald's has been to cater for Hindus and Sikhs, who are traditionally vegetarian, and also Muslims who do not eat pork. Chicken burgers are served alongside the McVeggie and McSpicy Paneer (an Indian cheese patty). In 2012, McDonald's opened a vegetarian restaurant for Sikh pilgrims visiting Amritsar, home to the Golden Temple. The success of these glocalising strategies owes much to the local knowledge of Connaught Plaza Restaurants who work with McDonald's in a joint venture (see page 27). | MTV Networks use a 360-degree strategy involving 'full spectrum' marketing to maximise its global audience. It has grown over time through acquisitions and mergers with existing companies, as well as by setting up new regional service providers such as MTV Base (available since 2005 to around 50 million viewers across 48 countries in sub-Saharan Africa via satellite). In 2008, new channel MTV Arabia began broadcasting to Egypt, Saudi Arabia and Dubai. Two-thirds of the Arab world is younger than 30 – and many of these young people are fans of cutting-edge music, especially hip-hop, which MTV Arabia now specialises in broadcasting. | *Roadside Romeo* (2008) was the first film that Disney made inside India. It was aimed at local audiences and used home-grown animation put together by Tata Elxsi's Visual Computing Labs (VCL) unit. A co-production with India's Yash Raj studios, this film tells the story of a dog living in Mumbai. Disney acquired Marvel in 2009, gaining the rights to superhero characters that have sometimes been glocalised. *Spider-man: India* is an example. In a story made for Indian children, Mumbai teenager Pavitr Prabhakar is given superpowers by a mystic being. The story is different from the version children in many other countries are familiar with. |
| 'We have to keep our ears to the ground to know what the local customer desires … [it's] key to our worldwide functioning.' (Amit Jatia, MD of McDonald's India West and South) | 'We will respect our audience's culture and upbringing without diluting the essence of MTV.' (Bhavneet Singh, MD of MTV Emerging Markets) | 'There is great interest and pride in local culture. Even though technology is breaking down borders, we're not seeing homogeneity of cultures.' (Bob Iger, Disney CEO) |

▲ **Table 2.5** Evidence suggesting that Western TNCs do not seek to impose their preconceived ideas on different places but are instead listening to what local audiences say they want

India's Bollywood film industry and Qatar's Al Jazeera TV channel. Children across the world are influenced by Japanese culture and ideas, notably Pokémon; Japanese influence is seen too in Lego's successful Ninjago franchise (see Figure 2.21). For decades, Asian technology TNCs such as Samsung and Sony have influenced how people consume music, TV and other entertainment.

Often, Anglo-American TNCs gain new ideas taken from other cultures, rather than attempting to supplant them. Global firms 'mine' different local territories to find new music, food or fashion

▲ **Figure 2.21** Japanese ideas exert a strong cultural influence on children in the UK

ideas that are fed back into Western markets. Japanese, Indian and Korean influences, among many others, increasingly drive innovation in the USA's creative industries. Contemporary film, music and food industries thrive by mixing together Asian, South American and African influences with European and American ideas. This is far from the idea of the West imposing cultural change on the rest of the world.

Movements of people also help create a melting pot of global culture. Each year, growing numbers of Chinese and Indian tourists spread their own culture and ideas while travelling worldwide (China now generates the highest volume of international tourism expenditure of any country in the world). Meanwhile, migration has turned most of the world's largest cities into places where many different cultures and ideas mingle together. US cities are home to a mix of people descended from Italy, Greece, Scandinavia, Scotland, Ireland, Mexico, Cuba, India, Pakistan, Vietnam, Korea and many other countries. In fact, one view of US culture sees it as an inclusive and adaptable entity that is continually modified over time by the arrival of new ethnic groups from all over the world.

Finally, never forget that Western culture and ideas have sometimes been violently resisted. The **war on terror** which began with the attack on the US World Trade Center in 2001 (see Figure 2.22) has sometimes been portrayed as a 'clash of civilisations'. Since then, the USA and its allies have been engaged in an ongoing struggle with militant groups such as al-Qaeda and Daesh (ISIS), including 'home-grown' terror incidents such as the attack on Westminster Bridge in 2017 and Nice, France, in 2016 (see Table 2.6). During this time, militant Islamist forces have used social media to portray their actions as a form of resistance against Western culture, religion and ideas. Chapter 5 returns to some of these themes.

▲ **Figure 2.22** Moments after the World Trade Center terror attack in 2001 (when two hijacked planes were flown into the buildings)

| Belgium | Suicide bombings in Brussels Airport resulted in 35 deaths and more than 300 wounded. |
|---|---|
| France | 84 people were killed and 100 more injured in the southern French city of Nice after a lorry was deliberately driven into a crowd celebrating Bastille Day. |
| Indonesia | In Jakarta, 2 people were killed and 24 injured in a terrorist attack which had been orchestrated and financed from Syria. |
| Libya | Militants detonated a truck bomb at a police training camp in Zliten, killing more than 50 people. |
| Pakistan | In a Taliban attack, 22 people were killed at Bacha Khan University. |
| Somalia | In the El Adde attack, Al-Shabaab terrorists attacked an African Union army base, killing 63 people. |
| USA | In a mass shooting, 49 people were killed and 53 injured at a nightclub in Orlando, Florida. |
| UK | 22 people were killed by a suicide bomber at Manchester Arena during an Ariana Grande concert. |

▲ **Table 2.6** Selected worldwide terror attacks in 2016 – 2017

# Reaching an evidenced conclusion

To some extent we can agree on the existence of a global culture which is rooted in the ideas, institutions and industries of Western countries, especially the USA. In the past, Western religion, cultures and languages were imposed on large parts of the world, and English, Spanish and French remain widely spoken languages. Critics of Western countries and companies point to 'one-way' processes of cultural transmission which still operate today, particularly the way some Anglo-American TNCs apparently hold sway over the entire planet's food, fashion and entertainment choices. These powerful companies use their considerable marketing expertise to recruit new customers in emerging economies like Brazil and India. Older and less commercial forms of leisure, play and entertainment are giving way to consumer aspirations in these and other countries. As a result, so the argument goes, local cultures are threatened or disappearing. With the death in 2010 of the last speaker of Bo, an ancient language of the Andaman Islands, India lost an irreplaceable part of its heritage.

The claim that local cultures and ideas remain resilient thanks to glocalisation is not entirely convincing. One can always argue that a McDonald's burger sold in India signifies the Americanisation of Asia irrespective of any 'tweaking' of the ingredients that occurs. Don't believe the 'hybrid' hype – a burger which pays 'lip service' to local traditions is nothing more than a 'Trojan horse' strategy. Similarly, some people may applaud the way MTV showcases local music in its different regional and national markets, but the bulk of its work involves promoting dominant Anglo-American music acts around the world. Any impression that local people are equal partners in a two-way 'cultural conversation' with media corporations such as MTV is probably false. This is because powerful Western TNCs often *do* work in coercive ways: they use clever advertising techniques to manipulate the desires and aspirations of people on a global scale (Noam Chomsky calls this 'the manufacturing of consent').

Yet we have now reached a point in history where the USA, UK and other Western countries must share the global stage with other powerful forces. A new multi-polar world is emerging (a theme which Chapter 6 returns to). The influence of China and India (soon to be the world's most populous state) will only keep growing. The diverse nations of Africa are projected to be the home of 4 billion people by 2100, and this will certainly affect how global culture and ideas are shaped in the twenty-first century. Finally, world events continue to remind us of the persisting way in which religion shapes personal and national identities and values around the world – sometimes in ways that are strongly opposed to Western ideas and culture.

 **KEY TERMS**

**Cultural traits** Culture can be broken down into individual component parts, such as the clothing people wear or their language. Each component is called a 'cultural trait'.

**Americanisation** The imposition and adoption of US cultural traits and values at a global scale.

**Westernisation** The imposition and adoption of a combination of mainly European and North American cultural traits and values at a global scale.

**Cultural homogeneity** At a local scale, this means most people share the same cultural traits, such as language, ethnicity and how they dress. At a global scale, it means that different places are losing their uniqueness and becoming increasingly similar to one another.

**Cultural hybridity** When a new culture develops whose traits combine two or more different sets of influences.

**War on terror** The ongoing campaign by the USA and its allies to counter international terrorism. It began as a response to al-Qaeda's attacks on the US World Trade Center and Pentagon in September 2001

# Chapter summary

✔ National governments can still control much of what crosses their borders despite claims we live in a 'borderless' or 'shrinking' world. Governments use a range of strategies including free-trade zones and low rates of corporation tax to attract global flows of capital.

✔ The attitudes of national governments to migration vary in time and space; some countries have far stricter controls on movements of people than others. Similarly, there is a spectrum of government attitudes towards the free movement of data across national boundaries.

✔ Intergovernmental organisations and agreements, including trade blocs, form an essential part of the architecture of global systems. The Bretton Woods Institutions have helped to produce and reproduce a neoliberal hegemony which the USA and its Western allies were originally instrumental in shaping. However, the power of institutions such as the IMF and World Bank is increasingly challenged by funds and banks managed by the governments of China and other emerging economies.

✔ Various international agreements exist that aim to regulate and manage global flows of people, including the freedom of movement granted to EU citizens. Since its inception, the UN has worked to protect the rights of refugees and deal with the consequences of forced migration.

✔ Superpower states have disproportionate influence over the way global systems operate. They try to establish economic and political frameworks that benefit themselves, using hard and soft power to regulate and reproduce the global systems they have helped make. Several states operate large sovereign wealth funds which generate enormous global capital flows.

✔ Western countries, especially the USA, have played a dominant role in the evolution of global systems, especially when it comes to the spread of culture and ideas. In the future, however, globalisation may increasingly be driven by non-Western influences too.

## Refresher questions

1 What is meant by the following geographical terms? Special economic zone; tax haven; transfer pricing; neoliberalism.

2 Using examples, outline ways in which national governments can help their countries attract more foreign investment.

3 Using examples, explain the reasons why a country's government might try to protect its businesses from foreign takeovers.

4 Using examples, explain why some countries have stricter migration rules than others.

5 Explain the role played by the Bretton Woods Institutions in the growth of global systems.

6 Outline the benefits which a country could gain from joining a trade bloc.

7 Using examples, outline ways in which the volume of international migration can be affected by international agreements.

8 What is meant by the following geographical terms? Superpower; hegemony; neo-colonial; cultural imperialism.

9 Using examples, explain how some countries use sovereign wealth funds to gain global power and influence.

10 Using examples, outline different ways in which Western countries have influenced culture and flows of ideas at a global scale.

## Discussion activities

1 If you held the political power needed to make change, what corporation tax rate would you set in the UK? Do you favour a high or low rate? A low rate may help attract foreign TNCs but could lead to less money being raised in taxes for spending on the NHS and good causes. Discuss possible arguments that can be applied to the decision-making process.

2 Discuss the extent to which global data flows (i) can be controlled, and (ii) ought to be controlled. Should people be given the freedom to look at whatever they want online (a liberal view) or should the state regulate internet use, possibly resulting in a so-called 'splinternet'?

3 Do nation states have less power than they used to? Discuss this statement in small groups, applying knowledge you may have gained in other subjects such as GCSE or A-level History. Think very carefully about what is meant by 'power' before answering!

4 In pairs, assess the ways the UK has used hard and soft power to gain influence in the world, both in the present and the past. How important is the UK today as a political, economic, military and cultural force acting on global systems? What evidence can be used to support your view?

5 To what extent would you say has your own life has been 'Americanised'? As a whole class activity, think about the influence of US culture on the lives of people in the UK (consider music, TV viewing habits, clothing and the speech/idioms we use).

# FIELDWORK FOCUS

This chapter's themes include government rules on migration and trade, and the global influence of powerful countries. There are some potentially interesting opportunities for an A-level independent investigation here. As with all global systems topics, you will need to think carefully about what kinds of primary data you might collect.

A *Using a mixture of questionnaire data and secondary sources to investigate how government migration policies have helped influence the character of a local neighbourhood.* Interviews could be conducted with business owners in an ethnically-diverse neighbourhood or town centre. The sample would need to be designed thoughtfully, in order to include only people who were born in other countries and migrated as adults to the UK. Questions might focus on the migration process, when it took place, and the ease with which the interviewees were able to gain UK residence rights and/or citizenship. Secondary research could be used to put the interview findings in context (for example, ONS information about source regions and sizes for past migration flows).

B *Investigating how a city's sports team has been changed by overseas investment flows, and the impact of these changes on local communities.* Many UK football teams are supported by overseas investment flows. Some teams are funded by billionaires and others by overseas sovereign wealth funds. Many teams' players were born overseas and have migrated to the UK. How have local communities in the UK been affected by the 'globalisation' and changing identity of their football teams? People's views and attitudes could be studied using primary data drawn from interviews or focus groups. It could be interesting to compare the perspectives of older and younger team supporters.

# Further reading

Barnett, C., Robinson, J. and Rose, G. (2008) *Geographies of Globalisation: A Demanding World*. London: Sage.

Driscoll, D. (1996) The IMF and the World Bank: How do they differ? Available at: https://www.imf.org/external/pubs/ft/exrp/differ/differ.pdf.

Herman, E. and Chomsky, N. (1988) *Manufacturing Consent: The Political Economy of the Mass Media*. New York: Pantheon Books.

Herod, A. (2009) *Geographies of Globalization*. Oxford: Wiley-Blackwell.

Jones, A. (2006) *The Dictionary of Globalization*. Cambridge: Polity.

Nye, J. (2005) *Soft Power: The Means to Success in World Politics*. New York: PublicAffairs.

# CHAPTER 3

# Global interdependence

The countries, places and people that are connected together within global systems are often described as being interdependent, meaning they have become mutually reliant on one another. Using examples of trade relationships and migration, this chapter:

- critically analyses the concept of interdependence
- investigates the extent of global economic interdependence resulting from trade and investment flows
- explores ways in which migration processes can make countries and communities more interdependent
- evaluates the extent to which interdependence creates more benefits than costs for different places.

## KEY CONCEPTS

**Global interconnectivity** All of the varied economic, social, political, cultural and environmental linkages between people, places and environments that make up global systems.

**Global interdependence** The idea that many states, societies and businesses have become mutually dependent on one another due to the complex ways in which global systems have evolved over time. There are economic, social, political and environmental dimensions of interdependence.

**World region** A large area of the world made up of multiple countries. For example, we might describe East Africa or Western Europe as world regions (the word 'region' is also used at an entirely different scale to mean part of a country, such as the English Lake District or Greater London).

**Risk** A real or perceived threat against any aspect of life. The growth of global systems has increased the exposure of people, companies and states to a range of physical, economic, political and technological risks.

## ① The concept of interdependence

▶ *In what different ways have the world's places and people become interdependent?*

The component parts of any system are, by definition, directly or indirectly interconnected. Things which are interdependent are more than just interconnected, however. They are *mutually reliant on one another* too.

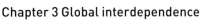

Figure 3.1 provides examples of how geographers view this important distinction between interconnectivity and interdependence.

| Interconnected (but *not* interdependent) things |
| --- |
| ■ In *coastal landscape systems*, inputs of eroded cliff materials provide the inputs that help build beach deposits. But while the beach might not exist without inputs from the cliff, the reverse is not true. The beach and cliff are not truly interdependent things. |
| ■ In *human geography*, there may be a causal link between the site of a place and the industries that grow there, for example mines are built where coal is found. However, the coal does not depend on the mines for its formation: they are not interdependent things. |
| **Interconnected *and* interdependent things** |
| ■ In *ecosystems*, vegetation biomass depends on a healthy soil for continued nutrient uptake. Equally, a healthy soil requires inputs of nutrients returned from the biomass via the litter store. Vegetation and soils are each dependent on the other for their health and survival. |

▲ **Figure 3.1** Establishing the difference between interconnectivity and interdependence

## Interdependence and global systems studies

In A-level geography, the concept of interdependence needs to be analysed carefully as part of Global Systems studies. To 'analyse' something means to break it down into component parts, and there are four commonly recognised deconstructed dimensions of interdependence: economic, social, political and environmental (see Figure 3.2).

Interdependence (in all its varied forms) is an emergent property of globalisation and the heightened complexity of global systems. Increased connectivity between places – in turn creating relations of mutual reliance – is driven by forces you will be familiar with from reading Chapters 1 and 2.

- Advancements in information and communications technology (ICT) have enabled the growth of extensive global production networks. Manufactured items ranging in size from phones to aeroplanes are assembled using components sourced from hundreds of different destinations. The headquarters, back offices and call centres of banks and other services are often scattered across continents while remaining networked together by ICT.
- Trading agreements and treaties between states have led to lower import tariffs, further encouraging partnerships between producers and suppliers in different places. According to economic theory, trade blocs also foster interdependence among their members as a result of market-place competition, as follows.
  1  In an expanded and borderless free market (such as that found within the EU), weak companies may fail while strong ones prosper.

**Economic interdependence**

- The growth of TNCs accelerates cross-border mutual exchanges of money and merchandise.
- Flows of migrant workers are often essential for a country's economic success; in return, migrants send home remittances to the source country.
- In an interconnected world, many countries are no longer self-sufficient in vital commodities such as food, energy or water; there are widespread relations of mutual reliance.

**Social interdependence**

- International migration builds extensive family networks across national borders. This can strengthen friendships between countries (e.g. India and the UK).
- Remittances sent home by migrants to a source country generate as much as 40% of GDP for some poorer states (e.g. Tajikistan). This helps families fund education and health. In turn, the migrants are often contributing to the health and education of the host country, e.g. NHS workers.

**Political interdependence**

- The growth of trade blocs (e.g. EU, ASEAN) has created complex international and global governance structures which countries co-create and are in turn legally bound by.
- The governments of some pairs or groups of countries historically depend on one another for support in times of economic or political crisis; for example, UK and US political leaders often refer to the two countries' so-called 'special relationship'.
- The World Bank, IMF and WTO work globally to harmonise national economic rules.

**Environmental interdependence**

- There has never been a greater need to co-operate on global environmental threats such as climate change, ocean pollution and biodiversity loss. Because all countries depend on these global commons, there is a mutually-felt need to share responsibility and help mitigate the risks.
- Agreements and treaties only work if a critical mass of countries sign up; all states depend on others to build a consensus (e.g. the 2015 Paris Agreement).

▲ **Figure 3.2** Four strands of interdependence in global systems: economic, social, political and environmental

🔑 **KEY TERM**

**Global commons** Global resources so large in scale that they lie outside of the political reach of any one state and, in theory, are open to all of humanity to utilise. International law identifies four global commons: the oceans, the atmosphere, Antarctica and outer space.

2 One member state's car makers could close because of superior foreign imports, yet its banks thrive in the large single market which now exists; meanwhile, another state may experience the exact opposite.

3 The populations of both countries can still buy cars and banking services, but they are now dependent on one another for their needs to be met.

- Continued global development and population growth have implications for environmental independence. The world's population keeps increasing in size and affluence, and its demands on global commons – including the oceans and atmosphere – rises accordingly (see Figure 3.3). We may already have passed planetary thresholds beyond which irreparable harm is now unavoidable, to the detriment of all. Environmental interdependence is a global phenomenon – and our future survival as a species may hinge on this vital understanding.

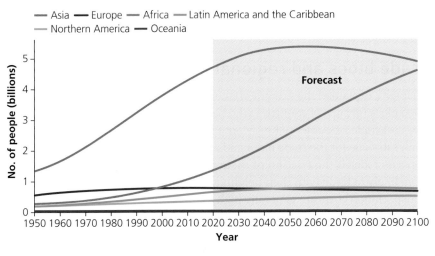

◀ **Figure 3.3** World population growth is forecast to continue until at least 2090, mainly on account of Africa. There are implications for global environmental interdependence: for the Earth to remain habitable, states must work together to find solutions for mutually shared resource challenges

Pearson Edexcel

AQA

OCR

WJEC/Eduqas

## Contested views about interdependence

Interdependence is an idea that often prompts strong debate. First, consider the proposition that the world economy is a single functioning global system composed of multiple interdependent parts and players. At face value, this sounds a sensible and interesting thing to say. But does the statement also imply – wrongly, perhaps – that the world economy is a mutually beneficial structure which everyone is happy to be part of? Oxfam reported recently that the world's eight richest billionaires – all of whom are men – collectively own the same wealth as the planet's 3.8 billion poorest people. In other words, human systems built on interdependent relationships often have outcomes that are extremely unfair. In human geography, it is therefore good practice to reflect critically on the link between interdependence and (in)equality.

- For example, think about the interdependence of employers and workers in economic systems at any scale, from small businesses to the largest TNCs. For centuries, philosophers have held opposing views about employer–worker interdependence.
- One perspective is that all employees benefit when a government allows employers to keep the lion's share of company profits. This is because there is an incentive for the employers to focus on running their businesses successfully (which means the workers keep their jobs).
- The opposing view is that profits should be shared more evenly between employers and workers because theirs is an interdependent relationship. In 2018, Amazon CEO Jeff Bezos became the world's richest person, with an estimated wealth in excess of US$100 billion. Bezos depends on his relatively poorly paid employees to make Amazon successful and they depend on him for work. But would you describe this as a 'mutually' beneficial relationship?

## KEY TERMS

**Equity** When everyone participating in a particular scenario or situation has been treated in fair and just ways.

**Friction of distance** A key geographical concept concerning the impediment to movement that occurs because people and places are spatially separated. The greater the separation, the greater is the friction of distance, meaning it's harder for things to flow to far places than to near places.

The linked themes of interdependence and equity are returned to in the final section of this chapter (see page 100).

# Trade blocs and regional interdependence

Chapter 2 explained the growth of intergovernmental trade agreements and trade blocs (groups of countries who have signed up to the same free or low-cost trade agreement). Key economic principles and theories provide the main rationale for the trade blocs and agreements shown in Figure 3.4. One important idea is the friction of distance: countries are, in principle, far more likely to trade with their neighbours than with a distant country if transportation costs are generally high. Supporting this, 65 per cent of exports in Europe are intra-regional, meaning that most European countries trade with other European countries. Similarly, neighbouring Singapore and Malaysia are close trading partners; so too are the USA and Canada.

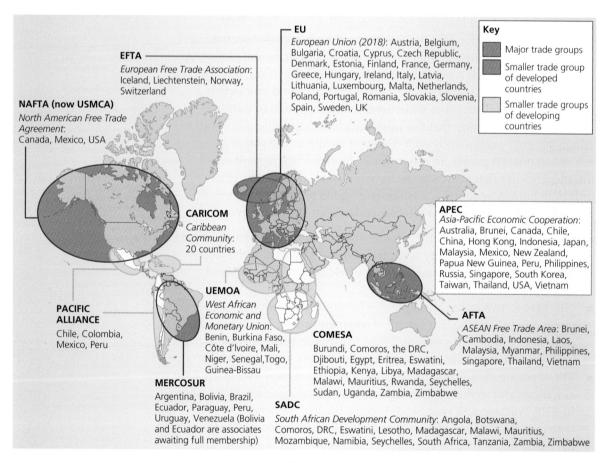

▲ **Figure 3.4** Trade blocs and agreements have fostered interdependence within and between different world regions

A second idea called comparative advantage helps explain enthusiasm for regional trading agreements. This principle states that if a particular country has the climate, resources or human skills that favour a particular kind of economic activity, then it stands to reason that country should specialise in those activities.

- Countries attempting autarkic development – meaning they 'go it alone' and try to produce everything they need by themselves – may fail to do so either cost-effectively or competently.
- A regional trade agreement fosters interdependence wherein each country specialises in whatever trade it does best while abandoning less profitable or successful ventures.
- Following the removal or reduction of import tariffs as part of a trading agreement, there is 'frictionless' movement of products and services across borders. Theoretically, this arrangement provides consumers throughout the trading area with competitively priced high-quality goods and services (see Figures 3.5 and 3.6).

**KEY TERMS**

**Comparative advantage**
The principle that countries should specialise in producing and exporting only the goods or services that they can produce at a lower relative cost than other countries.

**Autarkic development**
Progress that happens without any outside help. It is uncommon historically and impossible to imagine in the modern world. Even North Korea, which is portrayed as an isolated state, is supported by China and has dealings with other countries too.

Pearson Edexcel | AQA | OCR | WJEC/Eduqas

**Pre-trade bloc**

**Trade bloc (with customs union)**

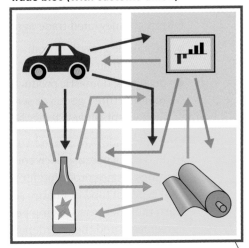

**Common external tariff wall**

All four countries produce their own goods and services across all four sectors (vehicles, textiles, financial services and wine); trade between one another at first is non-existent. As a result, output is limited in all cases and product costs are high for consumers. In some cases the goods and services may well be of poor quality, possibly due to inefficiencies resulting from a lack of high-calibre physical or human resources.

A trade agreement is now in place. National borders are rendered permeable for trade flows through removal of tariffs. Each country discovers it has a competitive advantage in one of the four sectors; these firms emerge as leaders in an enlarged market. Scales of production increase and costs of goods and services for consumers fall. A common external tariff may protect these 'winners' from overseas imports.

▲ **Figure 3.5** Trade specialisation within a regional trade bloc can be explained using the principles of (i) comparative advantage and (ii) economy of scale

▲ **Figure 3.6** Airtel Africa is a mobile phone company headquartered in Kenya whose expansion into 17 African states has been helped by the existence of the COMESA trade bloc

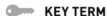
- A successful business based in one country may acquire similar industries in neighbouring countries and integrate them into its own expanding production network, or smaller firms within a trade bloc can merge to form a much larger TNC, thereby making their operations more cost-effective. Unilever and Royal Dutch Shell are products of European corporate mergers (Unilever was originally a small Dutch margarine manufacturer but has grown into an enormous conglomerate by acquiring over 400 other European brands spanning food, drink and household items).

The third important concept is called **economy of scale**. The enlarged market in a trade bloc increases demand for the goods and services of successful TNCs, thereby raising the volume of production while also lowering manufacturing costs *per unit*. These savings can be passed onto consumers through lower prices. Sales are therefore likely to rise even more for the most successful firms because the prices of their goods and services have now fallen – this process may keep repeating itself, creating a positive feedback effect (see page 37). Citizens of member states should find they can buy more with the same wages.

This all results in increased interdependence. Over time, each country within an integrated trade agreement will, in theory, experience:

- economic growth overall – on account of *economic success through specialisation*
- declining domestic productive capacity for certain types of food, manufactured goods or services – because other countries in the agreement *do these things better*.

One downside is increased vulnerability to any sustained interruption to trade. This is because each country has become more dependent on imports of things it once produced for itself but no longer does – possibly including vital types of food or energy supplies. It therefore comes as no surprise to learn that there are varying perspectives on the costs and benefits of trade agreements and the interdependence they bring. Moreover, the sheer complexity of cross-border investment flows can make it hard to deliver a valid and reliable cost–benefit analysis.

In North America, Mexico enjoys a comparative advantage over the USA as a manufacturing site on account of much lower wages. This has led many US companies to relocate their factories and operations across the border; examples include General Motors, General Electric and Nike. These companies have built a spatial division of labour which optimises its use of both countries' human resources. Goods are manufactured cheaply in branch plants in Mexico called maquiladoras, but are designed and marketed by highly skilled professionals in the USA. On the one hand, jobs have been lost in the USA. But on the other hand, US citizens can purchase goods made in Mexico costing less than when they were made in the USA. These issues and debates are returned to later in this chapter.

# ANALYSIS AND INTERPRETATION

Study Figure 3.7, which shows the value of export flows leaving the USA and Mexico and directed towards other NAFTA members in 2016.

**Value of all exports, 2016, in US$ billions**

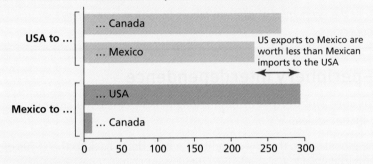

▲ **Figure 3.7** Trade relations between the USA and Mexico within NAFTA in 2016

(a) Analyse the pattern of exports shown in Figure 3.7.

### GUIDANCE

This is a relatively straightforward task involving simple data manipulation. You can measure the values shown and calculate the total value of exports for the USA and Mexico. You can also compare both countries' trade with Canada. There are good opportunities here to practise using phrases like 'almost twice as much' or 'over 20 times greater'. Language like this helps to convey the scale of any differences shown.

(b) Assess the extent to which Figure 3.7 demonstrates an interdependent relationship between the USA and Mexico.

### GUIDANCE

Used in conjunction with data like these, the command phrase 'assess the extent' ideally requires an answer that offers a critical appraisal of the information. You might assess its strengths and weaknesses, or comment on what has been included *but also what has been omitted*. Several possible themes can be developed around an assessment of how far Figure 3.7 demonstrates US–Mexican interdependence:

- Each country exports a large value of merchandise to the other, suggesting they are dependent on one another's industrial output.
- However, there is a value gap of around US$70 billion, suggesting an asymmetric relationship (the USA depends more on Mexico than vice versa).
- Moreover, we don't know anything about this trade relationship beyond dollar values. Perhaps the USA is exporting things that Mexico desperately needs and cannot get elsewhere (such as equipment to help build power stations or other vital infrastructure), whereas Mexico could be exporting mainly consumer items (TVs, fridges) that the USA could survive without or source from elsewhere. This would also represent asymmetric interdependence.

# 2 Global economic interdependence and supply chains

▶ *Why are the economies of different places becoming increasingly interdependent?*

## Core–periphery interdependence

The terms core and periphery have a long history of use in Geography, at varying spatial scales.

● At the global scale, there is a tradition which dates back to the 1970s of viewing the world as a single interdependent system comprised of a core, periphery and semi-periphery.
● Core–periphery interdependencies can also be identified and analysed at national and international scales.

Various core–periphery models were devised many years ago by Gunnar Myrdal (1957), Albert Hirschman (1958) and John Friedmann (1966). Even today, they remain a useful starting point for the investigation of economic interdependence, especially in relation to trade or migration flows.

● Core regions – at a global or more localised scale – are areas that enjoy cumulative growth processes fuelled by flows of raw materials, migrants and entrepreneurial talent from surrounding peripheral areas. This process of uneven development – the spatial movement and concentration of physical and human resources into a core – is called backwash.
● In time, growth is predicted to spread into the periphery as a result of market recompense for raw materials, the diffusion of innovations from core regions, government intervention and other beneficial interactions (see Figure 3.8). In Friedmann's account, a more complex core–periphery system develops over time as a result of this spreading wealth effect, which he termed 'trickle-down'. At first, secondary cores materialise in the periphery; later, a functionally interdependent system of linked core regions evolves.

The same optimistic trickle-down view of spatial economics has been embraced by successive US and UK governments, along with Bretton Woods monetary policy-makers, since the 1980s (see Chapter 2, page 43).

## World systems theory

Immanuel Wallerstein developed his world systems theory in the 1970s (see Figure 3.9). He argued that the world is divided into:

- *core regions* – North American and western European states, Australia, New Zealand, Japan and South Korea
- *semi-periphery regions* – the emerging economies of Latin America and Asia, including India and China, along with parts of northern and southern Africa and most of the Middle East
- *peripheral regions* – sub-Saharan Africa and some other parts of the developing world.

The relationships Wallerstein analysed are interdependent ones within a single capitalist world economy.

- Core countries and their companies get large returns on the foreign investment they make in semi-periphery countries.
- Periphery regions provide raw materials (food, energy) needed to supply the manufacturing industry in semi-periphery regions and consumers living in core regions.

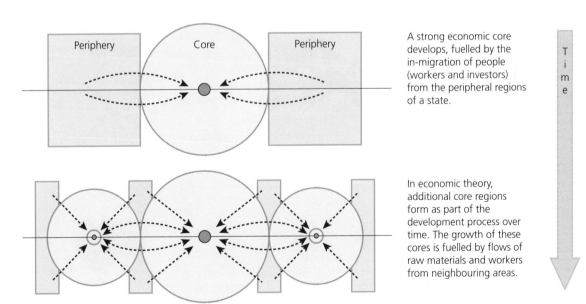

A strong economic core develops, fuelled by the in-migration of people (workers and investors) from the peripheral regions of a state.

In economic theory, additional core regions form as part of the development process over time. The growth of these cores is fuelled by flows of raw materials and workers from neighbouring areas.

▲ **Figure 3.8** The core–periphery model (Friedmann's version is shown here) can be applied at varying geographic scales to help us understand the way different countries, or parts of them, are parts of one interdependent system linked by reciprocal flows of merchandise, money, services, people and ideas

## Interdependence and equity

According to world systems theory, countries do not develop autonomously but are shaped over time by their interrelations with others, sometimes in positive ways. Gradually, the interdependence Wallerstein described has allowed semi-periphery states to emerge as a significant economic and political force. However, peripheral sub-Saharan countries have yet to benefit substantially from world trading relationships. Fundamentally, Wallerstein's system is composed of interdependent places and people; *but the profits generated by this system are not shared out equitably among the participants.*

The same prickly issues affect all variants of core–periphery theory (at whatever scale). There can be an implied mutuality of benefits arising from interdependent relations between different places, but this paints a disingenuous picture of global and national wealth distributions. By some measures, the world has become a staggeringly unequal place as a result of core–periphery flows. Globalisation has channelled money into the hands and pockets of a global elite who are no longer spatially contained within the borders of a mere handful of countries. Chapters 4 and 5 explore these issues of development, inequality and injustice in greater detail.

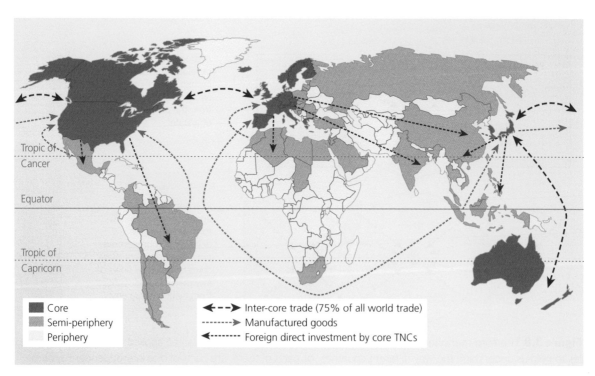

▲ **Figure 3.9** Wallerstein's model shows the growth of functional interdependence over time

# Global production networks and supply-chain interdependence

Chapter 1 (pages 26–27) explored the range of investment and marketing strategies used by TNCs to build their global businesses. Offshoring, outsourcing, mergers, acquisitions, joint ventures, glocalisation: all can be found in 'the TNC tool kit'. Often, an extensive global production network (GPN) – comprising both offshored and outsourced operations – will support the production of one single item.

Within GPN supply chains, outsourcing contractors often outsource in turn to other companies. The result is a linked chain of 'tiered' suppliers. Figure 3.10 shows how Apple has outsourced construction of its iPhone to a first-tier supplier, the Taiwanese company Foxconn. In turn, two hundred second-tier suppliers provide the parts Foxconn needs. Lianjian Technology is a third-tier Chinese company that delivers parts to the Korean firm, Wintek, which makes the iPhone touchscreen. In 2011, it emerged that Lianjian Technology workers had been exposed to dangerous chemicals; this generated bad publicity for Apple.

Many TNCs lack direct contact with their suppliers' suppliers and only inspect working conditions at the factories of their first- or second-tier suppliers. Increasingly, there is pressure for TNCs to show greater social responsibility by looking deeper into their supply chains for signs of worker exploitation (see also Figure 3.23, page 105).

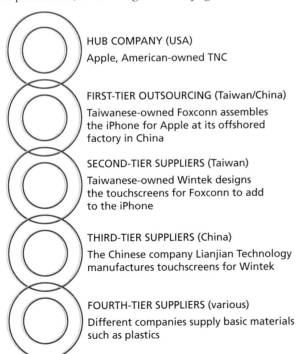

HUB COMPANY (USA)
Apple, American-owned TNC

FIRST-TIER OUTSOURCING (Taiwan/China)
Taiwanese-owned Foxconn assembles the iPhone for Apple at its offshored factory in China

SECOND-TIER SUPPLIERS (Taiwan)
Taiwanese-owned Wintek designs the touchscreens for Foxconn to add to the iPhone

THIRD-TIER SUPPLIERS (China)
The Chinese company Lianjian Technology manufactures touchscreens for Wintek

FOURTH-TIER SUPPLIERS (various)
Different companies supply basic materials such as plastics

▲ **Figure 3.10** A small part of the tiered supply chain for the iPhone, represented here as a linked chain of interdependent producers of parts and merchandise

## Supply chains supporting the UK car industry

A typical car made in the UK may use 30,000 first-tier components imported from around 15–20 different countries, mostly within the EU. Each first-tier component is composed of up to 30 second-tier sub-components, many of which will come from even further afield, for example China or Malaysia.

- With so many multiple upstream and downstream linkages in the supply chain, it quickly becomes impossible to pin down where a car like the Land Rover Discovery (see Figure 3.11) is really 'made'.
- The ownership structure of the Jaguar Land Rover (JLR) company reflects another kind of economic interdependence. The firm is based in Castle Bromwich and Halewood in the UK but was acquired in 2008 by India's Tata Motors. The survival of a great British brand therefore depended on investment from India. In turn, one of emerging India's most successful TNCs relies on investments such as JLR to help it gain a foothold in a well-established world market like the UK.

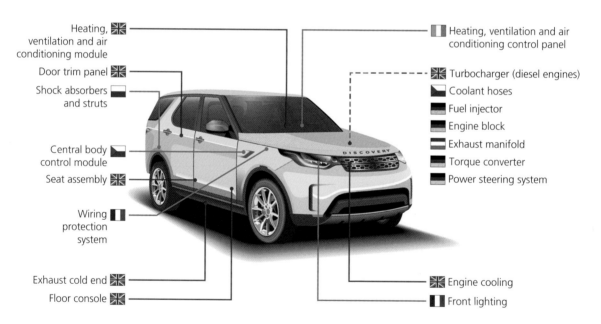

▲ **Figure 3.11** The Land Rover Discovery and the global production network that supports its construction. *Adapted from FT Graphic; Source: IHS Markit*

# CONTEMPORARY CASE STUDY: THE GLOBAL FINANCIAL CRISIS

In 2008, world money markets suffered a major shock that became known as the global financial crisis (GFC) or 'global credit crunch'. The interdependent nature of the global economy ensured that this disaster spread instantly and extensively, annihilating wealth on an unprecedented scale. In the twelve months after September 2008, world GDP fell for the first time since 1945. Globalisation was discovered to have a 'reverse gear'. This was not the first post-war global economic disaster; precursors include the major OPEC oil crisis of the 1970s and more recently a whole string of hiccups, such as the 'dot com' crash of 2001. However, the scale and magnitude of the economic harm done by the GFC was unprecedented. World trade fell at roughly twice the rate experienced during the Great Depression of the 1930s. Several major financial institutions failed and a near-total collapse of global economic confidence occurred, with share prices tumbling across all stock exchanges. Many national economies entered a period of recession (see Figure 3.12).

The GFC originated in US and EU financial markets, where sales of high-risk services and products triggered the bankruptcy or near-collapse of several leading banks and investment companies. One view is that blame for the GFC lies with people working in the financial sector, especially New York and London bankers who were all too ready to exploit the openness of the market and become involved in reckless speculation. Weak regulation of the market was a result of key government decisions (see page 43). Politicians deregulated banks in the 1990s, thereby allowing them to trade the financial risk of 'toxic' debt on to third parties via products known as bonds (see Figure 3.13).

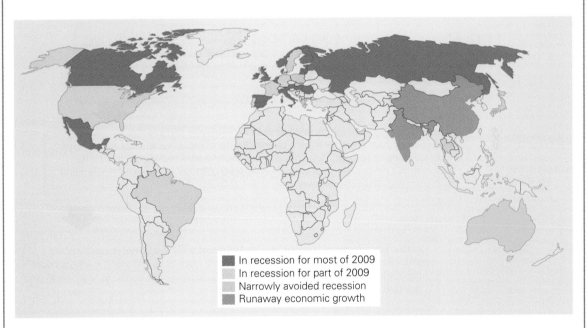

In recession for most of 2009
In recession for part of 2009
Narrowly avoided recession
Runaway economic growth

▲ **Figure 3.12** The interdependence of modern states meant that the negative effects of the GFC spread rapidly. In 2009, one year after the crisis began, large numbers of countries experienced negative economic growth on account of a problem which first developed in US money markets

According to some economists, the fundamentally unbalanced yet interdependent relationship between the USA and China was the most important root factor affecting all that happened. In the run-up to the GFC, China had developed an enormous trade surplus. This had left the Chinese state with a giant savings glut held in US dollars which subsequently depressed borrowing costs (interest rates) for US and European banking customers, in turn fuelling a cheap loans and spending bonanza. US–Chinese financial interdependence therefore underpinned the GFC (see Figure 3.14).

Before 2008, housing salespeople working on commission pushed up sales of sub-prime property mortgages in the USA (by offering large housing loans to poor people irrespective of their ability to pay back the debt). In the UK, lenders like Northern Rock also adopted a highly relaxed attitude towards lending credit (allowing people to borrow more money than their house was worth).

Banks turned the housing debt owed to them into bonds – a financial resource that could, in turn, be traded with other firms. This was so successful that it fostered a widespread sense of security in the banking industry. Any risk was perceived as well spread because so many organisations were investing in the same bonds market. This encouraged even riskier lending.

But as defaults started to rise on US sub-prime mortgages (because of rising unemployment), house prices fell. Bond values likewise shrank. In a culture of fear, confidence plummeted and many banks faced sudden bankruptcy after finding their assets were worth far less than previously calculated.

In September 2008, investment bank Lehman Brothers filed for bankruptcy. Following its multi-billion dollar collapse, facts emerged showing that bank had lent 35 times more money than it held in assets (meaning it only owned US$1 out of every US$35 registered in its accounts). This excessive leverage was unsustainable.

Sudden evaporation of financial liquidity left global trade freefalling. Households and corporations postponed major spending decisions (such as buying a car), leaving product manufacturers with warehouses full of unsold goods and often unable to find cash to pay their staff wages. G7 economies shrank in size.

▲ **Figure 3.13** The GFC originated in sales of high-risk financial products in high-income countries

## Consequences of the GFC for global systems and interdependent local places

The Asian Development Bank estimated that financial assets worldwide shed more than US$50 trillion in value during 2008 (a figure of the same order as total annual global economic output). These events represented a 'crisis of globalisation' – the GFC was fostered and transmitted by the rapid and deep financial integration of national economies. Extensive supply chains and production networks built up over decades left the global economy highly vulnerable to any fall-off in demand. Looking back at Figure 3.11, if demand for new cars were to ever fall suddenly again (as happened in 2009), then all of JLR's many first-tier and second-tier suppliers would also lose business. Interconnected and interdependent firms are vulnerable to a 'domino effect' that can lead to the disaggregation of entire industrial 'clusters' during a downturn.

According to the IMF, governments in North America and Europe spent $US9 trillion dollars in the first few months of the crisis as they scrambled to support 'systematically important' institutions. The UK government spent almost US$1 trillion protecting Lloyds TSB and the Royal Bank of Scotland (RBS) from bankruptcy, leading in turn to 'austerity measures' affecting all UK citizens, such as reduced spending on higher education and the arts.

Many millions of people lost their jobs worldwide during the GFC. The World Bank estimated that 90 million people living in Africa, Asia and Latin America were pushed back into extreme poverty. Table 3.1 shows how global flows – and some of the places these flows link together – were affected by events. Figure 3.15 focuses on countries that were most exposed to the GFC on account of unusually high levels of interdependence with other states and societies. Chapter 6 explores some of the longer-term impacts of the GFC and the austerity measures that followed.

| Migration flows | ■ An estimated half of the UK's eastern European migrants returned home temporarily or permanently soon after the start of the GFC (many returned later).<br>■ As many as 20 million Chinese workers lost their work in export processing zones and returned home to the countryside. Around 9000 export factories in the Pearl River Delta (Dongguan, Shenzhen, Guangzhou) shut for several months, causing massive temporary job losses. |
| --- | --- |
| Merchandise flows | ■ Instances of trade protectionism tripled during 2008 and 2009. Notably, the USA imposed a 35 per cent tariff on Chinese tyres, sparking a major trade dispute between the two superpowers.<br>■ South Africa lost 50 per cent of its iron ore export trade with Europe and Japan during this period. |
| Financial flows | ■ Indian workers flocked home from Dubai as construction projects ground to a halt in 2008. As a result, financial remittances plummeted too, falling in value by billions of dollars.<br>■ Total world value of FDI fell from US$1.7 trillion in 2007 to US$1.2 trillion in 2008 as falling demand for goods and services led TNCs to suspend expansion plans for new markets. |
| Tourism flows | ■ Visitor numbers to Thailand fell by 20 per cent. Many other countries recorded a fall-off, too.<br>■ Internal tourist flows rose for many countries as people opted for a 'staycation' instead of a foreign holiday. In the UK, many more people stayed at home and took holidays locally in Scotland or Cornwall instead of overseas. |
| Information flows | ■ This was one area that was unaffected. Increased membership of global virtual groups like Facebook and Twitter were the globalisation 'success stories' of 2008 and 2009.<br>■ 2009 was also the year that SEACOM completed its 17,000 km submarine fibre-optic cable connection linking East Africa to global networks via India and Europe. |

▲ **Table 3.1** How global flows were affected by the GFC in 2008–09 (when globalisation went into 'reverse gear')

USA
World's largest superpower
1 Mass sales of Chinese goods to customers in the USA and EU
2 China's firms and central state rake in enormous global profits
3 China builds up a money surplus, which it converts into US dollars
4 Chinese dollar assets bankroll lending at the heart of the credit crunch
China
New superpower in ascendance

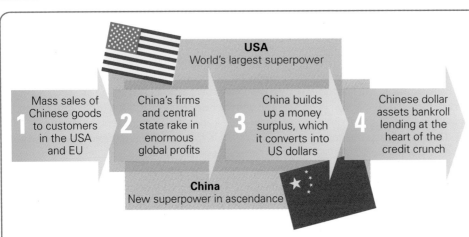

▲ **Figure 3.14** The financial interdependence of the USA and China was an important contributing factor to the GFC

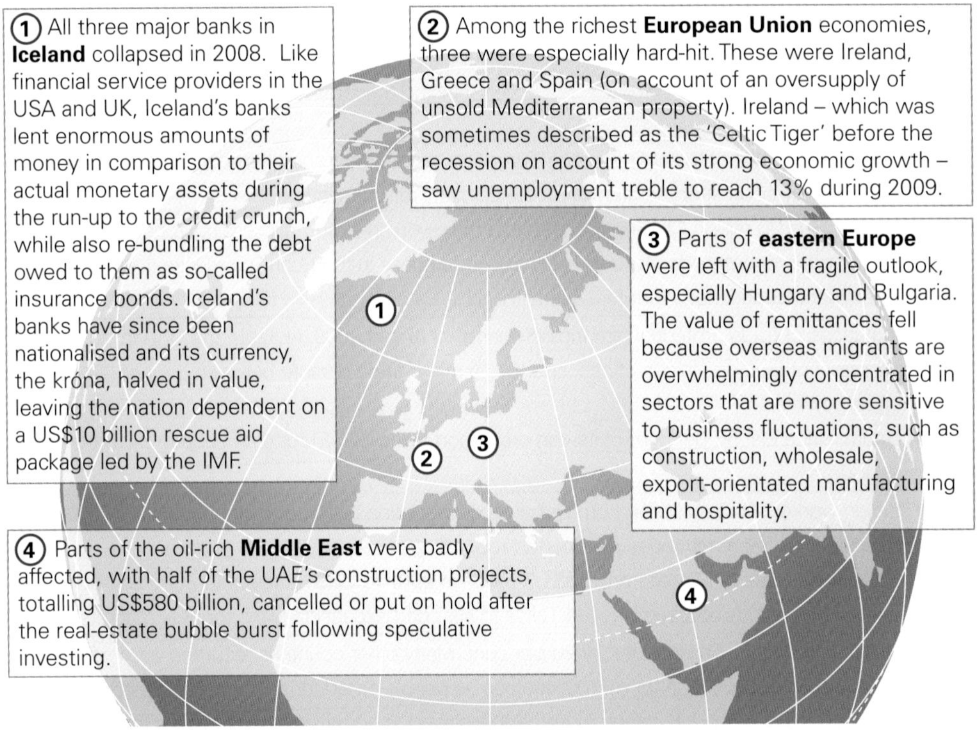

① All three major banks in **Iceland** collapsed in 2008. Like financial service providers in the USA and UK, Iceland's banks lent enormous amounts of money in comparison to their actual monetary assets during the run-up to the credit crunch, while also re-bundling the debt owed to them as so-called insurance bonds. Iceland's banks have since been nationalised and its currency, the krόna, halved in value, leaving the nation dependent on a US$10 billion rescue aid package led by the IMF.
② Among the richest **European Union** economies, three were especially hard-hit. These were Ireland, Greece and Spain (on account of an oversupply of unsold Mediterranean property). Ireland – which was sometimes described as the 'Celtic Tiger' before the recession on account of its strong economic growth – saw unemployment treble to reach 13% during 2009.
③ Parts of **eastern Europe** were left with a fragile outlook, especially Hungary and Bulgaria. The value of remittances fell because overseas migrants are overwhelmingly concentrated in sectors that are more sensitive to business fluctuations, such as construction, wholesale, export-orientated manufacturing and hospitality.
④ Parts of the oil-rich **Middle East** were badly affected, with half of the UAE's construction projects, totalling US$580 billion, cancelled or put on hold after the real-estate bubble burst following speculative investing.
①
②
③
④

▲ **Figure 3.15** States that suffered greatly during the GFC. Can you see how global interdependence had exposed them to an especially high level of risk?

Migration and interdependence

▶ *How has international migration led to different countries becoming increasingly interdependent?*

## Core–periphery migration patterns and processes

Chapter 1 outlined the importance of migration and remittance flows within global systems (see page 10). Economic migration is, in theory, mutually beneficial for source and host countries alike. Earlier in this chapter (see page 87) we explored the workings of core–periphery models. At a global scale, interdependent countries become linked together in economically beneficial ways by (i) backwash flows including economic migration and (ii) spread effects (or trickle-down), which include remittances. For example:

- Over 2 million Indian migrants live in the UAE, making up 30 per cent of the total population. Many live in Abu Dhabi and Dubai. An estimated US\$15 billion is returned to India annually as remittances. Most migrants work in transport, construction and manufacturing industries. Around one-fifth are professionals working in service industries.
- Around 1.5 million migrants from the Philippines have arrived in Saudi Arabia since 1973 when rising oil prices first began to bring new wealth to the country. Some work in construction and transport industries, others as doctors and nurses in capital city Riyadh. US\$7 billion is returned to the Philippines annually as remittances. Reports of ill-treatment of some migrants suggests there is a human cost to interdependence, however.

Mexican migration into the USA gives another vivid example of interdependence, along with the costs and benefits it brings. In theory, US businesses enjoy the fruits of cheap labour while remittances sent to Mexico help fund the social development of families. But there is more to migration than economics. The issue of legal and illegal migration across the Mexican border is a major policy issue which divides the public and politicians alike. While in office, President Obama called for work permits to be issued to many of the estimated 8 million unauthorised workers living in the USA. In contrast, while running for office President Trump demanded that a wall be built along the Mexican border. The migration issues which commonly divide US public opinion are shown in Table 3.2.

| Economic impacts | One view is that migrants are a vital part of the US economy's growth engine. From New York restaurant kitchens to California's vineyards, legal and illegal migrants work long hours for low pay. Citigroup Research suggests that two-thirds of US growth since 2011 is directly attributable to migration. This is because migration self-selects entrepreneurial people who are more likely to start their own businesses wherever they live. More than half of the USA's most valuable technology companies were founded by immigrants, as were 40 per cent of its 500 largest TNCs. But despite this evidence, high unemployment in some deindustrialised places has led to calls for immigration to be curbed. |
|---|---|
| National security | The terrorist attacks on the USA in 2001 ushered in an era of heightened security concerns. Support grew for the anti-immigration 'Tea Party' movement. In 2016, before he became US President, Donald Trump suggested that Muslims should be banned from entering the USA on the basis that global terror group Daesh (ISIS) pledges allegiance to Islam. Many people found this deeply offensive. Trump insisted he was simply thinking of ways to safeguard national security. |
| Demographic impacts | In the USA and other developed countries, youthful migration helps offset the costs of an ageing population. Yet the higher birth rates of some immigrant communities is changing the ethnic population composition of the USA. In 1950, 3 million US citizens were Hispanic. Today, the figure has reached 60 million. This is more than one-fifth of the population. |
| Cultural change | Migrants change places when they influence food, music and language. Hispanic population growth is affecting the content of US media as programmers and advertisers seek out a larger share of the audience by offering Spanish-language soap operas on platforms like Netflix. |

▲ **Table 3.2** Reasons for differing views about migration held by US citizens and organisations, irrespective of the mutual benefits that economic interdependence is meant to bring

### The European Union, migration and interdependence

At a larger continental scale, the EU's Schengen Agreement has accelerated the following backwash and trickle-down processes associated with international migration.

- The logic of the Schengen Agreement is rooted in an economic theory that views human beings as a resource that businesses need. People in eastern and southern Europe have therefore been allowed to move in large numbers to where most work is available, including France and Germany (see Figure 3.16 on page 98).
- Many economists believe this backwash process (to use Friedmann's terminology) works in everyone's interest. Migration is viewed as an efficient way of making sure that economic output is optimised for the EU as a whole.
- In turn, this provides EU governments with greater tax revenues to pay for services and infrastructure that are shared with all member states, including road-building projects, payments for farmers and grants for new businesses.

- It can therefore be argued that backwash losses suffered by some member states are balanced or even surpassed by a trickle-down of new investment. The result is a truly interdependent alliance of states.
- But critics of this neoliberal model argue that backwash migration costs for peripheral states in eastern Europe are really far greater than any trickle-down benefits they may gain. In reality, it is hard to either accept or reject the hypothesis due to the sheer complexity of the economic and demographic processes involved.

## CONTEMPORARY CASE STUDY: EASTERN EUROPEAN MIGRATION AND FINANCIAL INTERDEPENDENCE

Since joining the EU in 2004, eastern European countries have lost millions of workers but have made economic gains too. One effect of out-migration is new inflows of capital: the estimated 2–3 million people who have left Poland since 2004 have generated remittances valued at around 4 billion euros annually (amounting to around 60 billion between 2004 and 2019).

Additionally, TNCs headquartered in western Europe have relocated factories and offices to Poland, Hungary, the Czech Republic and Slovakia (also called the Visegrád 4, or V4). The attraction is a cheap, well-educated local workforce made up of those who have not migrated elsewhere. For example, in 2006 Nestlé announced the loss of 645 jobs at its factory in York (UK) and moved the production of Aero to the Czech Republic. Today, Poland is a popular destination for German and Asian investors, including South Korean technology giant LG. FDI into Poland equals around €6.5 billion annually.

Increasingly, though, FDI flows in both directions. As Poland's economy has grown (the World Bank now classes it as a high-income economy), some of its own companies have expanded abroad. Based in Gdansk in northern Poland, the retailing firm LPP is one of Poland's biggest companies – this TNC opened a store in London's Oxford Street in 2017.

As you can imagine, the balance of costs and benefits for Poland and the other V4 states is not easy to calculate accurately. Inward flows of investment and remittances must be offset against the loss of a sizeable proportion of young adults, including key workers (such as doctors

trained at the state's expense) and entrepreneurial talent capable of creating new ideas and wealth.

Recently, the shortfall in young workers on account of out-migration has begun to push up wages faster than productivity (with relatively few people unemployed, companies must compete for workers by offering higher pay).

To keep wages down (and profits up), companies in Poland are increasingly recruiting workers from the Philippines. Poland has also issued nearly 2 million short-term work permits to neighbouring Ukrainians. Public attitudes to immigration are not always positive in Poland though, and there are concerns about **community cohesion** should large numbers of migrants begin settling more permanently. An opposing view is that as the Philippines, Ukraine and Poland are all majority Catholic countries, the migrant workers may integrate well locally.

In the future, even more migrant workers will be needed. Eastern Europe has experienced a dramatic fall in fertility and, according to UN projections, the combined population of the V4 group will fall from 64 million today to around 55 million by 2050. Alternatively, industries may instead become more dependent upon inflows of new ideas and technology. There are already signs of significant investment in the latest artificial intelligence and industrial robots (see also page 190).

In conclusion, flows of people, money and technology have embedded the V4 states in regional and global systems in highly interconnected and interdependent ways (see Figure 3.18).

🔑 **KEY TERM**

**Community cohesion** When the diverse individuals and societies living in an area all share similar feelings of identity and belonging to that place.

► **Figure 3.16** Global flows of people and money make Poland a country whose economic and social sustainability is now reliant on a series of interdependent relationships involving other places and societies

**Polish out-migration**

Western European countries become dependent on Polish migrant workers.
Poland becomes dependent on their remittances.

**Labour shortages in Poland**

The loss of up to 3 million young people leaves gaps in Poland's workplace which cannot be filled easily.

**In-migration from Asia**

Poland grows dependent on migrant workers from the Phillipines.
The Phillipines becomes dependent on remittances from Poland.

## Diaspora communities and interdependence

The word 'diaspora' is used to describe the international or global dispersal of a particular nation's migrant population and their descendants. Global diasporas provide important contexts or frameworks for the study of international migration and the resulting linkages that are created between countries. Several famous diasporas are shown in Table 3.3. Other notable examples include the French, Italian, Mexican, Brazilian, Nigerian and Malay diasporas. The UK's 'Celtic fringes' have all birthed significant global diasporas despite these nations' relatively small population sizes. For instance, Ireland is home to just 4 million people, yet over 70 million individuals living worldwide claim Irish ancestry. In the USA alone, 30 million people believe they are tied to Irish bloodlines, following mass emigration from Ireland during the nineteenth and early twentieth centuries.

| | |
|---|---|
| **The Chinese diaspora** | Neighbouring Indonesia, Thailand and Malaysia, along with more distant states such as the UK and France, have significant Chinese populations. In many world cities, clearly delimited 'Chinatown' districts exist. A thousand years of sea-faring trade gives this diaspora a long history. The arrival of Chinese TNCs in Africa has brought further diaspora growth (see page 117). |
| **The Indian diaspora** | This is one of the world's largest, numbering 28 million in 2016. People of Indian citizenship or descent live in almost every part of the world. Important features of the distribution pattern are that it numbers more than 1 million in each of 11 countries. The largest concentrations are in the USA, UK, Malaysia, Sri Lanka, South Africa and the Middle East. |
| **The 'Black Atlantic' diaspora** | This has been described by writer Paul Gilroy as a 'transnational culture' built on the movements of people of African descent to Europe, the Caribbean and the Americas. A shared, spatially dislocated history of slavery originally helped shape this group's identity. Today, international connectivity and interdependence is maintained through migration, tourism and cultural exchanges across the Atlantic that are well exemplified by an international Black music scene that has given the world jazz, Jimi Hendrix, reggae, hip-hop and grime. |

▲ **Table 3.3** Examples of global diaspora populations

Former US presidents Barack Obama and John F. Kennedy fostered good diplomatic relations with the Republic of Ireland and the USA while in office (with both men being of Irish descent). The arrival of a large Korean diaspora population in the USA has deepened the country's friendship with South Korea.

Like Ireland, Scotland is a small country of only a few million residents yet has a diaspora numbering tens of millions. Online ancestry websites enable people living all over the world to trace their roots back to Scotland: this is another interesting way in which technology has influenced global migration. People who discover they have roots in another country may become more likely to consider moving there. GlobalScot is a website run by government-funded Scottish Enterprise that actively encourages members of the Scottish diaspora to network economically with one another. This means that interdependent relations between different segments of the diaspora population are actively constructed by Scotland's government.

▲ **Figure 3.17** This shop in Albany Road, Cardiff caters for the cultural and functional needs of the city's many eastern European diaspora communities

# ANALYSIS AND INTERPRETATION

Study Figure 3.18, which shows Chinese and Indian non-resident citizens living abroad in selected countries in 2011.

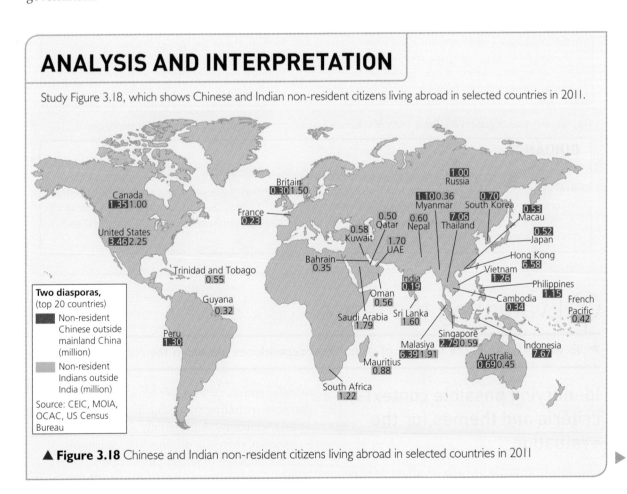

▲ **Figure 3.18** Chinese and Indian non-resident citizens living abroad in selected countries in 2011

**(a)** Analyse the distribution patterns for Indian and Chinese non-resident citizens living abroad.

> ### GUIDANCE
>
> There is a lot of information provided in Figure 3.18. When carrying out an analytical task like this, there is a skill in knowing *what to leave out* in the account you provide. It is important to not get too 'bogged down' in long lists of statistics. Instead, ask yourself: what is the 'big story' shown here? Are both patterns truly global? Are the two populations represented more in some continents than others? Are there any values which stand out as being particularly large and therefore significant? There are very high Chinese numbers in Thailand and Malaysia, for example. In order to provide a succinct analysis, you might consider working to a maximum word count of 150.

**(b)** Suggest why Figure 3.18 may only provide an incomplete view of the Indian and Chinese global diasporas.

> ### GUIDANCE
>
> There are several different ways of approaching this task. First, think about what the figure *actually* shows – which is Indian and Chinese non-residents citizens living abroad. These are people born in the two countries who have emigrated *recently*. However, the full diasporas of both countries additionally include people of Indian and Chinese descent whose parents or older ancestors migrated many years ago (and they are not shown in Figure 3.18). Second, there are data collection issues to consider. How accurate is the information and how has it been collected? Does it quickly go out of date, for example when people who have been away on business return home? Have estimates of illegal and unregistered migration been factored in?

**(c)** Explain how migration and the growth of diaspora populations can contribute to global interdependence.

> ### GUIDANCE
>
> This question provides plenty of opportunities for you to apply knowledge and understanding gained from reading Chapter 3. You might decide to adopt a global flows framework and explain, in turn, the importance of flows of people, money and ideas. Remember to maintain a strong focus throughout on the idea of *interdependence* (the mutual reliance that develops between places and societies) and not just connectivity.

#  Evaluating the issue

▶ *To what extent do the benefits of global interdependence outweigh the risks it creates?*

## Identifying possible contexts, criteria and themes for the evaluation

This debate prompts us to think about different possible *categories* of benefits and risks that global interdependence can create for a whole range of players – including states, businesses and ordinary citizens. Additionally, as this chapter has shown, there are different *dimensions* of interdependence to reflect on: economic, social, political and environmental. Each may bring its own benefits, risks and potential costs.

## 'Big picture' global benefits and risks

People who are strong advocates of globalisation – sometimes called hyperglobalisers – believe that it benefits humanity as a whole. According to this 'big picture' viewpoint:

- the feeling of interdependence gives rise to a shared set of values called global citizenship (see page 7)
- this increases the likelihood that the **Sustainable Development Goals (SDGs)** will be met: in an interconnected and interdependent world, the governments and citizens of states will collaborate to bring positive change because there is a shared sense that 'we're all in this together'. Or at least that's the theory.

There is a contrasting 'big picture' view, however, made up of various 'doomsday' scenarios. For example, the transport and communications technologies that support economic interdependence also risk allowing a biological or digital virus to spread worldwide, carried by global flows of people or data. These network risks are significant ones (see Figure 3.19).

## Benefits and risks in a local context

In addition to the macro-scale issues outlined above, we can evaluate the balance of benefits and risks for more specific local contexts, or groups of players.

- Before joining a trade bloc, any state's government (and electorate) must carefully weigh up the projected pros and cons of participating in 'frictionless' cross-border trade. EU states are additionally required to accept free movement of people as a condition of membership. A range of economic, social and cultural consequences – both positive and negative – may logically follow any country's decision to become an interdependent part of a larger bloc of states. Local places will undoubtedly change as a result. British people's lives have been shaped by the UK's EU membership over several decades, and not necessarily in ways which are to everyone's liking.
- We can view benefits, costs and risks from the point of view of large businesses too. Globalisation has allowed the world's most successful TNCs to thrive and in some cases attain trillion-dollar value (see page 30). But heightened interdependence makes companies vulnerable to new or increased physical, economic and political risks. Increasingly, companies are reassessing the overall cost-effectiveness of complex global production networks. The result is a new phenomenon called reshoring (see page 193).

## Thinking critically about complexity

Do the benefits of global interdependence outweigh the risks it creates? This is a provocative question to ask because in practice it's actually very hard to provide a definite answer with any real confidence. The sheer complexity of global systems makes it hard even for experts to

▲ **Figure 3.19** Globalisation and interdependence have meant that states, societies and businesses face an increasing number and range of hazard risks

quantify accurately what the net effects of greater interdependence have been (or will be) for the economies and populations of different countries. European and other regional supply chains and trading agreements are the product of evolution over many decades, and who can really know what the consequences of dismantling them would be? During this same time, feedback effects have sometimes intensified the challenges facing deindustrialised places while accelerating new growth and prosperity in other local contexts. Intangible and hard-to-measure variables also come into play, such as people's feelings about the cultural and social gains and losses that an interdependent and interconnected world brings.

## View 1: the benefits of global interdependence are worth the risks

As recently as the 1960s, a large part of the world's population remained disconnected from any global trade flows and instead engaged in local, rural subsistence activities. Times have changed since then; there are now more mobile phones than people on the planet and only a very small proportion of the world's population remains entirely disconnected from global systems. The majority of us are – to a greater or lesser extent – all parts of one great interdependent structure.

- Global agribusinesses have transformed rural economies throughout Asia, Africa and Latin America. They buy up land (see page 158) and may enrol local peasant farmers into their salaried cash-cropping workforce.
- These new farm workers are subsequently made dependent on TNCs like Cargill and Monsanto for their continued employment. In turn, the companies depend on their low-wage workforces for labour.
- Consumers in Europe and North America depend on the imported fruit, vegetables and

cereals which global systems provide them with – items which they might once have grown for themselves locally, but no longer do (see Figure 3.20).

▲ **Figure 3.20** In this UK supermarket, our dependency on other countries becomes clearly visible (tomatoes from Spain, avocados from Israel, champagne from France, etc.). In turn, what food or drink produced in the UK might you find in a supermarket in Spain or France?

One great upside to increased connectivity and interdependence – according to some longitudinal studies of international conflict – is that the world has, according to some measures, become more civilised and peaceful. There is a well-established literature that makes this case and includes work by psychologist Steven Pinker. He argues in his 2011 book *The Better Angels of Our Nature: Why Violence Has Declined* that interdependence has increased because of more stable world states, intergovernmental organisations, and complex trade and communications networks. As a result, the long-term trend is fewer deaths due to violence. Pinker cites data suggesting that fewer people die in international wars today than in the past (although figures may vary slightly from year to year, for instance during recent conflicts in Syria and the Middle East).

In an influential text written in the early 1990s, Thomas Friedman argued similarly that economic and political interdependence jointly help shape a more peaceful world. In his 'golden arches' theory of conflict prevention, Friedman asserted that two countries with McDonald's restaurants would never wage war because their economies had become interlinked (see Figure 3.21). Although the recent conflict between Russia and Ukraine has weakened this argument (both countries have McDonald's restaurants), it remains an idea worth exploring. Several years ago, Friedman updated his hypothesis, calling it instead the 'Dell Theory of Conflict Prevention'. No two countries that are part of the same global supply chain (such as that of the computer manufacturer Dell) will ever fight each other, he asserts. The economic risks of mutually assured destruction (both economies would suffer greatly if the global supply chain collapsed) are simply too high.

▲ **Figure 3.21** One argument is that these golden arches have helped bring world peace. What's your view, and why?

Then, of course, there is the work of the UN and its agencies to think about. There are many instances of the UN successfully fostering international co-operation, ranging from recognising the rights of refugees to tackling the problem of ozone depletion. Central to the UN's work is the promotion of globalisation, interdependence and international-mindedness as virtuous concepts: the idea being that societies who are mutually reliant on one another, like members of the same family, have the best incentive to work collaboratively to tackle shared challenges. The irony is, of course, that some of the most pressing issues of today – such as runaway carbon emissions and cyber-security threats – are created by the interdependent global systems which the UN values.

## Benefits for individual countries

The theoretical benefits of trade agreements for world regions such as Europe, North America or East Africa were outlined earlier in this chapter (according to the theory of comparative advantage, economic interdependence becomes a 'win-win' scenario for all countries involved following the removal of border tariffs to create a 'frictionless' free-trade area). There are plenty of data supporting the view that free trade has been highly beneficial at both the global and world-region scale. For example, world trade grew in value from around US$50 billion in 1950 to more than US$15 trillion in 2017. This fact is sometimes used to support the argument that globalisation is mutually beneficial for all parties and has translated into higher material welfare for most countries' populations. Spectacular increases over time in wellbeing for South Korea and China support the view that their integration into global economic systems – along with the interdependence this brings – has been most beneficial.

Yet free trade can create 'losers' as well as 'winners' in different localities, as any geographer will tell you. At face value, global and national GDP growth statistics suggest free trade and interdependence are good for all. But these headline aggregate figures don't give any indication of the complex pattern of geographic change occurring, i.e. the way that some local places have suffered greatly as a result of structural changes. Many blue-collar workers (see page 7) in the old industrialised heartlands of the USA and Europe have not had an easy ride. From the perspective of these populations and places, global shift has sometimes stripped away livelihoods and a sense of identity. It is these communities' disaffection with interdependence and globalisation that has helped fuel radical new political forces in recent years, including the UK Independence Party and the USA's Donald Trump (Figure 3.22).

▲ **Figure 3.22** Deindustrialisation has been a downside of interdependence for many places in developed countries such as the USA (Detroit pictured). Support for anti-globalisation movements may be high in these areas

## View 2: the risks and costs of interdependence are too great

The optimistic views of Pinker, Friedman and other hyperglobalists provide reassurance in an age of rapid change. But, their critics say, the case has been overstated. Later in this book, Chapter 6 explores contemporary trends in trade protectionism new nationalist (or 'nativist') political movements, the spread of AI and cyber-weapons, and rising geopolitical tensions between the USA, China and Russia. In their various ways, these are all outcomes of, or reactions against, globalisation and interdependence. It is increasingly hard to accept Friedman's argument that a new and durable era of global enlightenment is propped up by McDonald's golden arches.

Recent world events have shown us the serious risks created by a networked and interdependent world.

- *Financial risks.* The economic shock of the global financial crisis (see page 91) spread widely and deeply across global systems in ways which revealed the risks of interdependence. For example, when the UK entered recession in 2009 as a result of the GFC, many building projects were cancelled. The knock-on effect was that many migrants working in construction industries lost their jobs. They stopped sending remittances home; as a result, Estonia's economy shrank by 13 per cent. This highlights the challenges that accompany the benefits of interdependence.
- *Technological risks.* In 2017, a computer virus called WannaCry spread to 200,000 computers in offices, banks and oil companies around the world. Hospitals across the UK were forced to suspend some operations. This was symptomatic of the newfound vulnerability of societies everywhere arising from shared use of ICT networks with Microsoft, Apple or Android operating systems. Cyber-attack risks will continue to grow unless there is coordinated action at citizen, institutional, civil society, national and international scales to build resilience.
- *Health risks.* More people than ever before routinely travel long international distances for reasons of work or pleasure; the West African Ebola virus epidemic (2013–16) provided a worrying glimpse of the harm that a truly global pandemic could bring.

## Risks for local places

Significant local losses can arise because of interdependence. Global systems constantly create new jobs in some locations while annihilating employment elsewhere. This first became truly apparent in developed nations during the 1960s and 1970s when destructive waves of deindustrialisation tore through the industrial heartlands of Europe and North America.

- The UK's major cities were hard hit; Liverpool lost 200,000 jobs and almost half its population between 1951 and 1981. While the global shift of productive industries often benefited local communities in Asia and Latin America, it was accompanied by a steep decline elsewhere.

- One view is that the gains of the maquiladoras (page 84) came at the expense of US blue-collar workers' jobs. Mexico benefited from the 12 new maquiladoras established there by General Motors (G.M.) in 1987, for instance. However, 29,000 jobs were lost when G.M. closed 11 plants in the Midwest of the USA that same year. The scale of losses in the Midwest and Great Lakes was so great it gave rise to the pejorative term 'rust belt' during the 1980s.

However, there is also a longer-term view that deindustrialisation ultimately provided the impetus for the positive remaking and regeneration of many cities across the Western world, adding further complexity to the argument.

## Risks for TNCs

To judge by their recent behaviour, risks may increasingly outweigh the benefits of interdependence for some TNCs. Many firms have built extensive global production networks composed of offshoring and outsourcing relationships, as this chapter (page 89) and Chapter 1 (see page 26) explained. All of this was originally made possible by political decisions (tariff removal between countries) and technological innovation (intermodal container shipping and ICT networks that keep suppliers in constant contact with the hub companies they provide for).

But supply-chain operations are not risk-free. Table 3.4 provides details of global risk exposure for businesses during the last decade or so. As you can see, supply-chain interdependence greatly increases the vulnerability of companies to disrupted operations on account of (i) natural hazards and (ii) conflict and geopolitical hazard risks. Furthermore, there are reputational (moral hazard) risks to watch out for. Reports of supply-chain worker exploitation or modern slavery can be extremely harmful for big brands. For example, forced labour is endemic in West African cocoa production; and at least one well-known global chocolate brand has had its reputation tarnished as a result of reports that its supply chain uses child labour (Figure 3.23).

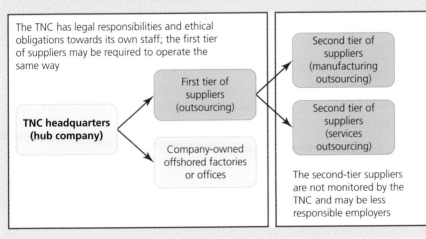

The TNC has legal responsibilities and ethical obligations towards its own staff; the first tier of suppliers may be required to operate the same way

TNC headquarters (hub company)

First tier of suppliers (outsourcing)

Company-owned offshored factories or offices

Second tier of suppliers (manufacturing outsourcing)

Second tier of suppliers (services outsourcing)

The second-tier suppliers are not monitored by the TNC and may be less responsible employers

◀ **Figure 3.23** One consequence of complex global production networks is the 'invisibility' of workers in lower-tier parts of the supply chain. If stories of exploitation surface, reputational harm may be done to the TNCs who do business with these lower-tier companies

| Geopolitical and conflict risks | *Political events can create supply-chain shocks for TNCs.*<br>■ US and EU oil companies operating in Russia were told by their governments to suspend any new operations there after sanctions were imposed following Russia's annexation of Crimea in 2014.<br>■ Many tourist and airline companies have lost custom as a result of conflicts and political instability: tourist revenues in Tunisia halved between 2010 and 2014 on account of Daesh (ISIS) terror attacks targeting holidaymakers.<br>■ The 'Arab Spring' wave of uprisings in North Africa in 2011 meant French companies such as France Télécom experienced service and supply-chain disruption across French-speaking North Africa (work stopped at the Teleperformance call centre in Tunis, for instance).<br>■ **Resource nationalism** can result in a TNC's overseas operations being seized by a state government (see Chapter 2, page 46). Regime change in a country may lead a TNC into conflict with new leadership, such as when socialist President Hugo Chávez seized control of ExxonMobil and ConocoPhillips operations in Venezuela. |
|---|---|
| Reputational (moral and ethical) hazard risks | *Unethical treatment of supply-chain workers and environments by outsourcing companies jeopardises the reputation of the TNCs who do business with them.*<br>■ In highly publicised 2010 court cases, European companies BP and Trafigura both tried to lay the blame for catastrophic environmental damage (in the Gulf of Mexico and Côte d'Ivoire, respectively) at the doors of subcontractors. Ignorance of harm done by a subcontractor to people or the environment rapidly translates into a 'moral' and reputational hazard situation for the hub company (see Figure 3.23).<br>■ Reports of child labour or worker suicides (China, 2010) can damage a company's reputation. A spate of fires in Dhaka during 2010 exposed clothing companies Gap and Hennes to brand association with poorly-monitored and unsafe factories owned by the sub-contracting firm Ha-meem Group (in the worst case, 26 workers died when fire exits were blocked). Similarly, the Rana Plaza disaster was a wake-up call for many TNCs (see page 151).<br>■ An emerging concern for some TNCs is discovering that merchandise produced by prisoners in Chinese labour camps has found its way into their supply chains. |
| Physical risks | *Natural hazards can disrupt supply chains unexpectedly.*<br>■ Millions of cubic metres of volcanic material known as tephra were ejected over Iceland when Eyjafjallajökull erupted in 2010. Flights were grounded for weeks after vast amounts of fine ash rose 9 km into the paths of jet aeroplanes. Kenyan farmers and flower suppliers went out of business because they couldn't get products to market in the UK (5000 supply-chain workers were temporarily laid off). Car production in Europe also ground to a halt as the flight ban starved firms like Nissan of key parts.<br>■ Japan's 2011 earthquake and tsunami highlighted the over-exposure of firms across the world to environmental supply-chain risk. US car makers Ford and Chrysler ceased production of red and black vehicles because the Merck factory that was the sole supplier of a vital metallic paint-pigment lay in the tsunami zone. |

▲ **Table 3.4** Recent examples of global risk exposure for TNCs arising from supply-chain interdependence

# View 3: the true benefits and risks are unclear

One common theme in UK and US current affairs is the new politicisation of interdependence. The EU aims to construct mutually dependent and beneficial relations among its member countries. Political interdependence, the argument goes, builds peace (while also putting EU states on an equal geopolitical footing with the USA, China and Russia). Economic interdependence is also meant to be a 'win-win' situation for all members, as explained above. But 'leavers' in the 2016 UK Brexit referendum refuted this. They claimed that quitting the EU would actually save the UK £350 million a week (see Figure 3.24). In turn, some 'remainers' accused the 'leavers' of lying and cited an alternative projection showing that a 'no deal' exit might reduce UK growth by eight per cent over 15 years.

Views differ too within the USA on the mutuality of the country's interdependence with Mexico within 1994's North American Free Trade Agreement (NAFTA) and its 2018 successor, the United States–Mexico–Canada Agreement (USMCA). Large volumes of merchandise and parts flow both ways across the US–Mexican border. Supporters of these agreements say free trade (i) has brought strong economic growth and emerging economy status to Mexico (while also discouraging illegal migration) and (ii) has provided the USA and Canada with a low-cost investment location and a new market for their exports. Like UK 'remainers', NAFTA advocates have data which support their own argument: one 2017 report calculated NAFTA had tripled cross-border merchandise and money flows, and added 0.5 per cent to the USA's GDP. In stark contrast to this, Donald Trump once called NAFTA 'the worst trade deal ever', blaming it for the loss of 600,000 US jobs.

Why do 'facts' about the EU and NAFTA differ so greatly according to who is telling the story? Why

▲ **Figure 3.24** In the 2016 EU referendum, some 'leavers' implied that £350 million would become available each week to spend on the nation's health instead

do interdependent geographical relationships generate such controversy? It is because of the complexity of the relationships under discussion. Vast interconnected supply chains have developed over time. Positive feedback effects often amplify particular employment gains and losses in different places. As a result, net impacts become ever harder to quantify. Moreover, the changes associated with trade agreements are concurrent with other forces, including rapid technological innovation, domestic political decisions and expanded trade with other countries (such as China). Different researchers may decide to factor in these other variables in their calculations, or not – perhaps because of their own political views on political interdependence and what they hope to show.

# Reaching an evidenced conclusion

The years 1980 to 2008 were a 'golden age' for globalisation, thanks to technology and falling barriers to overseas investment. For governments and businesses, the economic benefits of 'acting globally' seemed to outweigh any risks created by interdependence. However, more recent world events – such as the GFC and the 2011 Japanese tsunami – have prompted some TNCs to re-examine supply-chain hazards and in some cases contemplate adaption or mitigation strategies such as reshoring. Fear of future black swan events (extreme occurrences that bring disproportionate impacts) can lessen the day-to-day appeal of globalisation's supposed benefits.

We are still seeing plenty of fresh growth in global production networks though. Newer players such as India's Tata and China's SAIC aim to compete on an equal cost footing with longer-established American, European and Japanese rivals. Outsourcing and foreign investment strategies are already being applied by these rising stars, some of whom seem less concerned with avoiding reputational hazards in contrast with their Western competitors. Chinese industries are widely perceived to be relatively relaxed about possible moral risks when doing business in Myanmar and North Korea, for example (see also the account of the Belt and Road Initiative on page 66). In contrast, successful TNCs like Apple, Walmart and BP are working harder than ever before to reduce their global risk exposure by auditing the behaviour of their suppliers and subcontractors more carefully.

Tools exist for companies to design a risk roadmap (one that ideally identifies critical points of failure, such as a single component for which no alternative source exists). But the reality is that engaging beyond first-tier suppliers remains a challenge for many companies. Walmart has an estimated 60,000 suppliers providing it with the merchandise sold to half a billion customers each week, and many of these companies are, in turn, reliant on parts and ingredients from other firms. For anyone operating on this scale, the potential risks created by interdependence are hard to eliminate altogether.

In closing, interdependence is one of the most important specialised concepts for A-level Geography. Its use raises important questions about the true nature of spatial and social relations between people, places, objects and environments. We can also reflect usefully on the benefits or risks that arise from interdependent relations, and the extent to which these fair or unfair effects are mutually shared too.

## 🔑 KEY TERMS

**Sustainable Development Goals (SDGs)** The UN's 17 SDGs were introduced in 2015. They replace and extend the earlier Millennium Development Goals (MDGs) which were a set of targets agreed in 2000 by world leaders. Both the SDGs and earlier MDGs provide a 'roadmap' for human development by setting out priorities for action.

**International-mindedness** A way of thinking which is receptive to ideas from different countries and recognises that all people belong to a networked international community which is pluralistic, culturally diverse and meritocratic. It also involves an appreciation of the complexity of our world and our interactions with one another.

**Remaking** An umbrella term which brings together a range of regeneration, redevelopment and reimaging actions intended to provide assistance to deindustrialised or declining places.

**Regeneration** Large-scale forms of state intervention and private-sector inward investment which attempt to transform the fabric of place, often on a very large-scale. The aim is to attract new investment into an area while trying also to stimulate local enterprise.

**Resource nationalism** When state governments restrict exports to other countries in order to give their own domestic industries and consumers priority access to the national resources found within their borders.

**Black swan events** 'Unthinkable', high-impact, hard-to-predict and rare occurrences. They bring disproportionate impacts for affected people and places. This concept was developed by Nassim Nicholas Taleb.

## Chapter summary

✔ Interconnectivity and interdependence mean different things: when connected people and places become mutually reliant on one another, they also become interdependent. There are economic, social, political and environmental dimensions of interdependence.

✔ Regional trade agreements have fostered interdependence among groups of countries in different parts of the world. These create conditions that help successful TNCs to thrive; but member states may become more reliant on imported goods and services than they used to be.

✔ Interdependent core–periphery structures can be identified at local, national, international and global scales. Backwash and trickle-down flows link core and periphery regions together. However, views differ about the extent to which the benefits of interdependence are mutually shared between core and periphery regions.

✔ The growth of TNC supply chains and global production networks have contributed to global interdependence. The global financial crisis showed how vulnerability and risk have also spread globally because of this interdependence.

✔ Migration and remittances can be viewed as backwash and trickle-down flows within core–periphery systems. Over time, migration may give rise to the growth of diaspora communities that have an important role to play in making different places and countries more interdependent.

✔ Views differ on whether interdependence is a good thing at global or more local scales. Because global systems and flows are extremely complex, it is often difficult to quantify benefits, costs and risks. This is one of the reasons why there was a lack of agreement about whether or not the UK should remain part of the EU.

## Refresher questions

1 What is meant by the following geographical terms? Friction of distance; comparative advantage; autarkic development; spatial division of labour.

2 Using examples, outline the ways in which countries can become economically interdependent; socially interdependent; politically interdependent; environmentally interdependent.

3 Using examples, outline the advantages of trade bloc membership for member-state businesses.

4 Using examples, (i) outline what is meant by the terms core and periphery and (ii) explain how core and periphery areas are linked together as part of a single system by different flows.

5 Using examples and an annotated diagram, explain how supply chains are used to support the production of manufactured products.

6 Briefly explain the causes of the global financial crisis (GFC). Outline what the consequences of the GFC were for three different countries.

7 Using examples, explain how international migration can result in economic interdependence for source and host countries.

8 Using examples, explain why diaspora populations play an important role in global systems.

9 Analyse the extent of inequality within different interdependent relationships you have studied.

## Discussion activities

1. What kind of interdependent relationships exist within your school or college? For example, do different subject departments, such as Geography and Maths, ever work together in mutually supportive ways? Where else in everyday life can we find examples of interdependence among groups of people?

2. Working in small groups, discuss the extent to which different interdependent relationships you have studied are also fair or unfair relationships for the people and/or places involved. What data sources would you need to consult in order to assess the extent of any inequality in a relationship?

3. Working in pairs, research a trade bloc or agreement other than the EU or NAFTA. Find out who the members are and why they wanted to join. In what ways does your chosen trade bloc foster interdependence among its members?

When each pair has completed the research, findings can be discussed with the rest of the class, perhaps using a presentation.

4. In small groups, think critically about the supply chains supporting the making of merchandise you own. Questions to ask are: where are the items you buy, such as a phone or trainers, manufactured? What parts make up these items? Where might these parts have come from? There are tools online that can help you carry out research or find out more about the issues, for example: http://getstring3.com/projects/

5. The GFC happened in 2008, over ten years ago. But its effects continue to be felt in various ways in the UK because of long-term effects on pay levels and austerity measures affecting everything from the cost of attending university to support for the arts. In small groups or pairs, carry out further research into these issues or discuss your prior knowledge.

## FIELDWORK FOCUS

The concept of global interdependence could provide the basis for a fieldwork investigation. For example, a study might focus on the origin of goods sold in UK high street shops; one of the investigation's aims being to explore the extent to which consumers make purchasing decisions based on a sense of 'connection' with producers (the people who produce the food and goods they are buying).

- Attempting to audit the entire contents of a supermarket or grocery store to find out where things were made or grown is, of

course, an unrealistic goal. You might therefore want to sample only particular types of goods, for example focusing on fruit, vegetables and/or meat. These are all unprocessed items; the packaging will usually carry clear information about the region of origin (meaning there are opportunities to map your data, or represent it graphically).

- This work could be augmented with interviews focusing on how far the general public take into consideration where the food they buy comes from. Do some people like to buy or avoid products from particular countries, and why?

# Further reading

Baylis, J., Smith, S. and Owens, P. (2016) *The Globalization of World Politics: An Introduction to International Relations*. Oxford University Press.

Friedman, T. (2007) *The World is Flat: the Globalized World in the Twenty-First Century* 3rd ed. London: Penguin.

George, R. (2013) *Deep Sea and Foreign Going*. London: Portobello.

Gilroy, P. (1993) *The Black Atlantic*. London: Verso.

Held, D. and McGrew, A. (eds.) (2000) *The Global Transformations Reader: An Introduction to the Globalization Debate*. Cambridge: Polity Press.

Knox, P., Agnew, J. and McCarthy, L. (2008) *The Geography of the World Economy*. 5th ed. London: Arnold.

Murray, W. (2006) *Geographies of Globalization*. London: Routledge.

Pinker, S. (2011) *The Better Angels of Our Nature: The Decline of Violence in History and its Causes*. London: Penguin.

Taleb, N. N. (2011) *The Black Swan: The Impact of the Highly Improbable*. London: Penguin.

# CHAPTER 4

# Global development and inequality

There is evidence linking globalisation with global development. Many previously poor countries are now classed as emerging economies. However, the benefits of global economic growth have not been shared evenly between different places, societies and individuals. This chapter:

- investigates the links between globalisation, trade and global development
- explores the role of migration flows in the development process
- analyses the increasing importance of data flows for global growth and development
- evaluates the impact of global systems on patterns of inequality.

## KEY CONCEPTS

**Development** Human development generally means a society's economic progress accompanied by improving quality of life. A country's level of development is shown first by economic indicators of average national wealth and/or income, but it encompasses social and political criteria also.

**Inequality** The social and economic (income and/or wealth) disparities that exist both between and within different societies or groups of people. Inequalities at global, national and local scales can be decreased or increased by flows of trade, investment and migration. Inequality in a society may increase even when everyone's incomes are rising because of the disproportionately large gains made by the wealthiest individuals and elite social groups.

 # Global development, trade and investment

▶ *In what ways can trade and investment contribute to global development?*

Chapters 1–3 explored how global systems produce and reproduce certain relationships between people, places and environments. This chapter's focus is on how far these systems have promoted growth and development both globally and locally. Human development, like globalisation, is a multi-dimensional process which can be studied at varying geographic scales. There is a large area of overlap (see Figure 4.1) between development studies and global systems studies.

- Both topic areas are concerned with economic disparities and the factors that can reduce or reproduce them.

- The financial flows that allow global systems to operate also transfer wealth between places in ways which can narrow or exacerbate different kinds of **development gap**.

Global Systems

- Impacts of financial flows (trade, aid, loans and remittances)
- Global governance in support of sustainable development
- Globalisation of democratic norms and support for human rights
- Global action in support of improved health and education

Human development

# Human development and ways of measuring it

Figure 4.2 shows 'the development cable'. It presents the development process as a complex series of interlinked outcomes for people and places. Among many other things, it shows that in an economically developed society:

▲ **Figure 4.I** The overlap between global systems studies and development studies

- citizens enjoy health, long life and an education that meets their capacity for learning
- citizenship and human rights are more likely to be established and protected.

Figure 4.3 shows the social changes that sometimes follow when the world's poorest farmers receive a boost in earnings on account of their integration into global systems (in this example, flows of money have connected poor farmers in a local context with the international Fairtrade organisation). This brief illustration highlights how economic, social, cultural and political changes are all interlinked parts of the human development process.

🔑 **KEY TERM**

**Development gap** A term used to describe the polarisation of the world's population into 'haves' and 'have-nots'. It is usually measured in terms of economic and social development indicators. Development gaps exist both between and within states and societies.

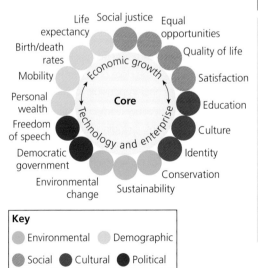

**Key**
- Environmental
- Demographic
- Social
- Cultural
- Political

▲ **Figure 4.2** The development cable

**Farming incomes start to rise**

**School attendance improves**

**Choice and opportunity**

**Fairtrade** (offers a higher guaranteed price for products)

**Microloans** (affordable credit that can help farmers grow a bigger crop and sell the surplus for cash)

**Increased pay** (due to large companies acting ethically and responsibly when they buy food and other goods from suppliers)

**More time** (children no longer have to work on the farm)

**More money** (families can afford to pay for school fees and school books)

**Less hunger** (children do not have to go to school hungry and can concentrate better)

**Poverty reduction** (by reducing poverty, a range of human needs are met, such as health and education)

**Life choices** (as they grow up, children can take the decision not to be farmers; they can choose other forms of work or study for a profession instead)

▲**Figure 4.3** Economic and social development linkages

## The validity and reliability of human development indicators

Table 4.1 explains how three important human development indicators are used and offers a brief evaluation of their validity and/or reliability.

| Indicator | Explanation | Evaluation |
|---|---|---|
| Income per capita | ■ The mean average income of a group of people is calculated by taking an aggregate source of income for a country, or smaller region, and dividing it by population size. This yields a crude average which can give a misleadingly high 'typical' figure if large numbers of high-earners inflate the mean.<br><br>■ Per capita gross domestic product (GDP) is a widely-used proxy for this. It is the final value of the output of goods and services inside a nation's borders (i.e. an estimate of the nation's income). A country's annual calculation includes the value added by locally-operated foreign-owned businesses.<br><br>■ The World Bank recently estimated global nominal GDP in 2014 at about US$78 trillion. Using this figure, can you make an estimate of average global GDP per capita? What other information do you need to find out? | ■ There is near-universal agreement that it is perfectly valid to study income levels as part of an enquiry into human development.<br><br>■ Recording GDP reliably is not always possible though. The earnings of every citizen and business need to be accounted for, including informal-sector work. Also, to make comparisons, each country's GDP is converted into US dollars. However, some data may become unreliable because of changes in currency exchange rates.<br><br>■ GDP data must be manipulated further to factor in the cost of living, known as purchasing power parity (PPP). In low-cost economies, where goods and services are relatively affordable, GDP size is increased by statisticians to reflect this (and vice versa). This is why sources often show two estimates of a country's GDP (Brazil had a 'nominal' GDP of US$1.9 trillion in 2018 and a 'PPP' GDP of US$3.4 trillion). |
| Human development index (HDI) | ■ The HDI is a composite measure that ranks countries according to economic criteria (GDP per capita, adjusted for PPP) and social criteria (life expectancy and literacy).<br><br>■ It was devised by the United Nations Development Programme (UNDP) and has been used in its current form since 2010.<br><br>■ The three 'ingredients' are processed to produce a number between 0 and 1. In 2018, Norway was ranked in first place (0.95) and Niger was ranked in last place (0.36). | ■ The three ingredients of the HDI – wealth, health and education – are widely regarded as valid indicators of development. All governments value wealth and health, while citizens' education supports these and other goals.<br><br>■ Literacy and life expectancy information is not always easy to record reliably. In recent years, millions of people have been displaced by human and/or physical-induced disasters such as conflict in Syria or drought in the Horn of Africa. This makes accurate HDI data near-impossible to collect. |
| Gender inequality index (GII) | ■ The GII is a composite index devised by the UN. It measures gender inequalities related to three aspects of social and economic development (see Figure 4.4).<br><br>■ Its ingredients are: reproductive health (measured by maternal mortality ratio and adolescent birth rates); empowerment (measured in part by parliamentary seats occupied by females); labour force participation (ratio of female and male populations in the workforce). | ■ Some states do not allow women to stand for election to parliament, including Kuwait. In Pakistan's Swat Valley, Taliban militia have burned down girls' schools. Cultures which do not support equal rights for women are unlikely to view the GII as a wholly valid development measure.<br><br>■ Collecting reliable data on labour force participation rates may be tricky due to the numbers of women who work in the informal sector or under 'zero-hours' contracts. |

▲ **Table 4.1** Evaluating the validity and reliability of different development indicators

Human development is measured in many different ways using both single and composite (combined) measures. When assessing the value of different measures, it is helpful to distinguish between issues of validity and reliability.

- For a measure to be *valid*, there should be broad agreement that it has relevance. Do you agree that political corruption should be used as a measure of development, for instance? Should gender equality be included as one of the most important aspects of human development, as the Millennium Development Goals did? Should we consider a country's commitment to sustainable development policies and climate change mitigation when attempting to analyse varying levels of human development?
- To be *reliable*, a measure must use trustworthy data. Do you think all countries' income and employment data are fully accurate, for instance? Might data measurement or comparability issues mean that we should sometimes question the reliability of different countries' estimates of their fertility, mortality or literacy rates? Should estimates of illegal global flows (see Chapter 5, page 154) be included in measurements of national income and wealth creation?

## KEY TERMS

**Millennium Development Goals** A set of interrelated global targets for poverty reduction and human development. They were introduced in 2000 at the UN Millennium Summit; their successor, the Sustainable Development Goals, followed in 2015.

**Sustainable development** A 'roadmap' for development that aims to ensure that the current generation of people should not damage the environment in ways that will threaten future generations' quality of life.

◀ **Figure 4.4** The status of women in society (including their voting and property ownership rights) is widely (but not yet universally) thought of as an important development measure

## Global trade patterns

Trade is the movement of goods and services from producers to consumers. It spans many different sectors of industry. Physical trade in goods includes movements of:

- primary industry products (food, energy and raw materials)
- manufactured items (ranging from processed food to electronics).

Overall, world trade is dominated by developed countries and several large emerging economies (EEs) including the BRIC group (the four large economies of Brazil, Russia, India and China). The following points provide

## KEY TERM

**Informal-sector work** Unofficial forms of employment that are not easily made subject to government regulation or taxation. Sometimes called 'the black economy' or 'cash in hand' work, informal employment may be the only kind of work that poorly skilled people can get.

**BRIC group** An acronym for Brazil, Russia, India and China. These four countries have large economies, large populations and each has showed a high growth rate in recent years. An annual summit meeting held with South Africa is called the BRICS summit.

an overview of global trade patterns with consideration to both production (origin) and consumption (market).

- The value of world trade and global GDP has risen by around two per cent annually since 1945 with the exception of 2008–09 when the global financial crisis (GFC) led to a brief fall in activity.
- Just ten nations, including China, the USA, Germany and Japan, account for more than half of all global trade.
- Around half of all trade originating in developed countries takes place with other developed countries (see Figure 4.5). This is because of the large numbers of affluent consumers and markets found in the world's wealthiest countries.
- Consumer markets have expanded in emerging economies as spending power has grown among their citizens. Middle-class higher-protein diets are characterised by greater consumption of meat and dairy. For example, China's annual meat consumption per capita rose from around 5 kg to 50 kg and Brazil's from 30 kg to 80 kg during the 20-year period 1990–2010.
- China remains the number one exporter of goods (valued at US$2.3 trillion in 2017) and thus a dominant influence on world trade. Indeed, a slowdown in the rate of Chinese growth since 2010 has been responsible for a 'cooling off' of the global economy as a whole. In particular, falling Chinese demand for imports of natural resources and oil has been financially harmful for some African exporters.

▶ **Figure 4.5** Global merchandise trade, 2015. Direction of movement is indicated by an arrow, while the width of the arrow is drawn proportional to the volume of movement. Using the information, can you estimate the value of the world's five largest trade flows?

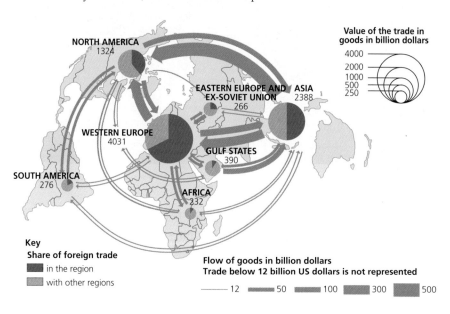

# CONTEMPORARY CASE STUDY: EMERGING CHINA AND THE GROWTH OF SOUTH–SOUTH TRADE

More than 40 years have passed since China opened its economy to overseas investors and embraced globalisation. In 1978, Deng Xiaoping initiated an era of 'reform and opening up' that involved free-market reforms and the welcoming of foreign investment (parallels can be drawn with the historical trajectory of Indonesia, shown previously on page 60). Most people then still lived in rural poverty in rural areas. But in the years that followed, agricultural communes were dismantled and farmers allowed to make a small profit for the first time. China's transformation into an urban, industrial country gained momentum, fuelled by the mass migration of 300 million rural migrants searching for a better life in cities. The country's economy and society developed at breakneck speed in two distinct phases.

1 Initially, urbanisation went hand-in-hand with the growth of low-wage factories, giving China the nickname the 'workshop of the world'. It became the top destination for outsourcing and inward investment from the world's largest TNCs into newly-established special economic zones (see page 43) in the Pearl River Delta, Shanghai and other coastal regions. Figure 4.6 shows the country's share of global trade rising rapidly by the mid-1990s. Workforce wages often remained very low, however, raising concerns about the ethics of outsourcing to China (see page 150).

2 Between 2010 and 2018, wages in China's manufacturing sector tripled, following worker protests and a shortage of younger labourers (the legacy of China's now-abandoned one-child rule for families). As a result, companies have sought to produce higher-value goods and much of China's trade has moved up the value chain, including more electronic goods and components such as electrical transformers (see Figure 4.7). Labour-intensive, cheaper-priced exports including 'budget' clothing increasingly originate from neighbouring countries like Vietnam instead of China. Another result of recent structural changes is the rapid emergence of an enormous Chinese middle class. This is helping a more balanced economy to develop that is no longer entirely reliant on foreign exports. Today, GDP growth also derives from increased private consumption of goods and services by China's increasingly affluent workforce.

Foreign investment remains important for Chinese trade and industry. In 2018, it accounted for around 45 per cent of the country's exports, including laptops and phones made for companies like Apple.

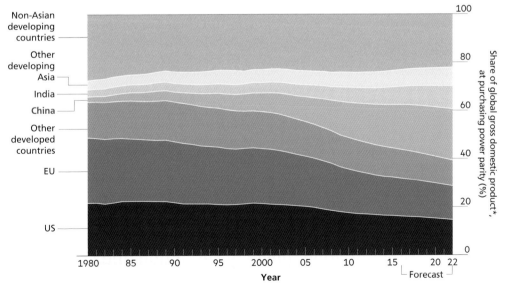

* Excludes Commonwealth of Independent States

▲ **Figure 4.6** China is an emerging economy (or economically developing country, EDC) whose share of global trade has expanded enormously since 1978 The global pattern of trade has changed markedly since 1980. In particular, China's share is now many times larger. *FT Graphic; Source IMF*

## China's role in south–south trade and migration flows

China's growth in turn has increasing influence on the economic development of the African continent. Flows of Chinese investment and migrant labour are directed towards, for example, South Africa, Ethiopia and Kenya. Between 2006 and 2016, Chinese imports to sub-Saharan Africa increased by nearly 250 per cent, reaching a value of US$170 million (this is 20 times higher than at the start of the millennium). Chinese companies sometimes invest directly in African countries too. This activity spans all sectors of industry, ranging from oil production in Nigeria and Sudan to tourism in Egypt and South Africa.

Accompanying these trade and money flows, over one million economic migrants have moved to Africa (around one-third live in South Africa, having travelled there from China's Fujian province). Some are entrepreneurs hoping to set up their own businesses. Many are contract workers serving Chinese state-owned companies such as Sinopec; around 250,000 belong to this category, including people who have migrated to help manage Belt and Road projects (see page 66).

Chinese companies have also invested significantly in South America, including US$20 million in Brazil during 2016 and 2017. Sinopec spent US$7 billion dollars acquiring almost half of Brazilian oil company Repsol in 2010, giving China an important stake in some of Brazil's recent offshore oil discoveries. Analysts say investment in Brazil is part of a state-led directive to ensure future food and energy security for China. Critics label such moves pejoratively as 'land grabs', but by all accounts the Brazilian government welcomes the investment. Relations between the two BRIC countries are increasingly interdependent.

It may be, however, that fewer migrants will move from China to African and South American countries in the future because rising wages have recently lessened the incentive for low-skilled Chinese workers to migrate elsewhere.

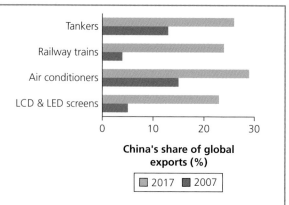

▲ **Figure 4.7** Since the early 2000s, China's export market has increasingly captured a greater share of high-value technology products

## A global success story?

In closing, China's story may lend support to the hyperglobal view that global-scale free trade can cure poverty. Yet we should be careful not to use China's growth as evidence to support the uncritical assertion that global systems *always* bring growth and development to countries. Not all developing countries that have opened their doors to global flows have fared nearly so well as China, whose story is unusual. Its government has not always played by WTO rules, and many world governments, including the USA, believe Chinese companies gain an unfair trading advantage through state subsidies. The US–China bilateral trade gap hit a record US$37 billion in China's favour in 2017, triggering the Trump administration's imposition of US tariffs on many Chinese goods in 2018 (see page 177).

Further analysis of China's growth trajectory appears throughout this book. Themes include:

- interdependence between China and the USA (see page 94)
- Chinese investment in the UK (see page 63)
- the human cost of China's 'economic miracle' (see page 150).

## Globalisation, poverty reduction and the new global middle class

Overall, the global economy has grown enormously since the mid-twentieth century, and far faster than the world's population (see Figure 4.8). One result is that nearly 1 billion people have been lifted out of absolute

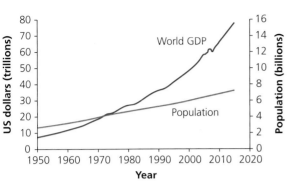

▲ **Figure 4.8** Comparing growth in world GDP and world population size, 1950–2015

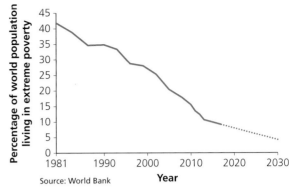

Source: World Bank

▲ **Figure 4.9** The falling share of the world's population living in extreme poverty (earning less than US$1.90 a day), 1981–2030 (projected)

poverty during the past couple of decades (see Figure 4.9). At a national scale, this is linked with the transition many countries have made from 'low-income' and 'developing' to 'middle-income' and 'emerging' status. Much of the growth has taken place in Asia and Latin America. Between 500 million and 600 million people (estimates vary) have been lifted out of poverty in China alone.

The World Bank is the most important source of information about **extreme poverty** today and sets the international poverty line. This measure was revised in 2015: a person is currently considered to live in extreme poverty if they are living on less than US$1.90 per day. According to this benchmark, the proportion of people in extreme poverty in developing countries (excluding China) fell from 40 per cent to 25 per cent between 1990 and 2010. When China is included, the figures are 46 per cent and 22 per cent – even more impressive. This shows that Chinese economic development has played a key role in helping global poverty-reduction targets to be met. In 2016, the number of people remaining in extreme poverty was estimated to have fallen below 800 million by the World Bank.

- This means that 90 per cent of the world's population now live above the extreme poverty line. A growing number belong to the new global middle class (see page 20): this means they earn or spend more than US$3650 a year, or US$10 a day.

- A further 2 billion people belong to a poorer group called the **fragile middle class**. They earn or spend between US$2 and US$10 a day: theirs is a precarious position and they might easily slide back into poverty should an economic crisis, conflict or major natural disaster affect the place where they live.

- However, poverty remains widespread in sub-Saharan Africa and some parts of southern Asia. The poverty rate in sub-Saharan Africa fell only 8 percentage points between 1981 and 2016.

**KEY TERMS**

**Extreme poverty** When a person's income is too low for basic human needs to be met, potentially resulting in hunger and homelessness.

**Fragile middle class** Globally, there are 2 billion people who have escaped poverty but have yet to join the so-called global middle class. Typically, they earn between US$2 and US$10 a day. The 'fragile middle' class is broadly similar to the idea of a global 'lower middle' class.

The reasons for these changes are complex but relate in part to globalisation and trade, in addition to work carried out by the UN and other international agencies. Richard Freeman of Harvard University has attributed some of the changes to a population dynamic he calls 'the great doubling': the global labour force doubled in size from 1.5 to 3 billion people when China, India and eastern Europe countries began to participate more fully in the world economy after a series of political changes in the 1980s. Another theory is that emerging economies have enjoyed the 'trickle-down' of scientific and medical know-how from Europe, North America and Japan.

In truth, there are many reasons for the broad 'headline' of global poverty reduction, and development trends in different countries should be analysed on a case-by-case basis. According to globalisation critics, it may also be the case that the 'success story' of poverty reduction has been overstated. Figure 4.10 shows how data can be used selectively to present a number of contrasting viewpoints on development and poverty in the world today.

| | | |
|---|---|---|
| In 2018, poverty in sub-Saharan Africa remains at nearly **50%** – no lower than in 1981; and population growth has meant the *number* of poor living there has doubled, from 200 million to about 400 million. | Poverty in East Asia fell from 80% of the population living below US$1.90 a day in 1981 to 18% in 2005. Much of the progress was in China, where **0.5 billion** people have been lifted out of extreme poverty. | Of the billion people worldwide who have escaped **US$1.90** a day absolute poverty since the 1980s, most would still be deemed as being very poor by European and North American standards. |
| The richest **eight** people alive in 2018 held personal wealth equivalent to that possessed by the poorest half of humanity, numbering **3.8 billion** people in total. | The global GDP of **US$80 trillion** is shared unevenly between rich and poor countries. It is also shared unevenly among rich and poor people living within different countries. | **800 million** people live on less than US$1.90 a day (the World Bank extreme poverty measure); and **2 billion** people get by on just US$2–10 a day (purchasing power equivalent). |
| Around 350 million Indians lived in poverty in 2017; yet 41 of the 1000 richest people on the planet were also Indian citizens – including **two** of the **top 50** earners. | The world's richest nations are also home to 100 million people who live below these places' official poverty line; including **43 million** US citizens (in 2015). | Between 1988 and 2011, the incomes of the poorest 10% increased by just US$65, while the incomes of the richest 1% grew by US$11,800 – **182 times as much**. |

▲ **Figure 4.10** Analysing evidence of poverty and inequality at different scales. Based on this evidence would you say global systems have, overall, brought greater growth and prosperity to most people – or not?

# CONTEMPORARY CASE STUDY: THE GLOBAL MIDDLE CLASS

Historically, the phrase 'middle class' came to describe a socio-economic group sandwiched between the workers (or 'working class') and the 'ruling class' in European countries. Today, the idea of a global middle class (GMC) is used to describe a growing mass of people who no longer experience the absolute poverty or low incomes still endured by almost 3 billion people globally. However, they may not have yet achieved the affluent lifestyles of 'the Western world' or Japan and South Korea.

Opinions vary on what exactly defines this 'middle class'. Essentially it describes those who are left with disposable income after essentials (shelter, heating, food) have all been paid for. At the bottom end of the middle-class income range, this might include someone who can afford to buy a non-essential can of Coca-Cola. At the upper end of the income range, it means having enough money to buy a fridge, phone or even a cheap car.

■ By one estimate, there are currently 2 billion people in the GMC, but this will grow to about 3.5 billion by 2020 as more of the low-income fragile middle class see their incomes rise (see Figure 4.11).

■ Asia is almost entirely responsible for this growth: its middle class is forecast to triple in size to 1.7 billion by 2020 and by 2030 Asia will be the home of 3 billion middle-class people (ten times more than North America).

■ There is also substantial middle-class growth in the rest of the emerging world (see Table 4.2). The

middle class in Latin America is expected to grow from 180 million to 310 million by 2030, led by Brazil. In Africa and the Middle East, it is projected to more than double, from 140 million to 340 million.

Remember, however, that estimating a national population's income or spending is not easy. Different countries use different currencies. Also, money goes further in some places than others. Income figures for different countries are therefore adjusted – sometimes crudely – to take into account purchasing power parity (PPP). For example, the average income in China is roughly US$8,000, but this becomes US$18,000 when adjusted for PPP (2018 data). It is therefore important to remember that some data used to report the new GMC phenomenon may not be entirely accurate, trustworthy or easily comparable.

| Country | Population (millions, 2017) | Middle-class credentials |
|---|---|---|
| Indonesia | 263 | The Indonesian middle class (people earning more than US$10 a day) is predicted to grow from 45 million in 2015 to 135 million by 2030. |
| Mexico | 129 | 70 per cent were middle-class in 2017, each spending over US$9000 annually. Mexico is the USA's second-largest export market. |
| India | 1339 | Only one-in-ten people were middle-class in 2017, but it could be one-in-five by 2025. India's retail market is worth almost US$1 trillion annually. |
| China | 1410 | One-in-three were middle-class in 2017. They spent around US$1.5 trillion, making China the world's largest market. By 2022, 550 million people will be middle-class. |
| Total | 3.35 billion | |

▲ **Table 4.2** Large emerging economies and their growing middle (consumer) class, 2017

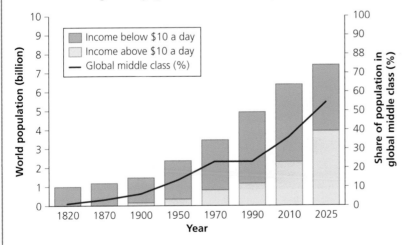

▲ **Figure 4.11** Actual and projected growth of the new GMC, 1820–2025

# ANALYSIS AND INTERPRETATION

Figure 4.12 is a representation of how the world's population changed between 1988 and 2011. For each world region, we are shown the changing distribution of wealth during this time period; this includes both the income range and the proportion of people earning different incomes.

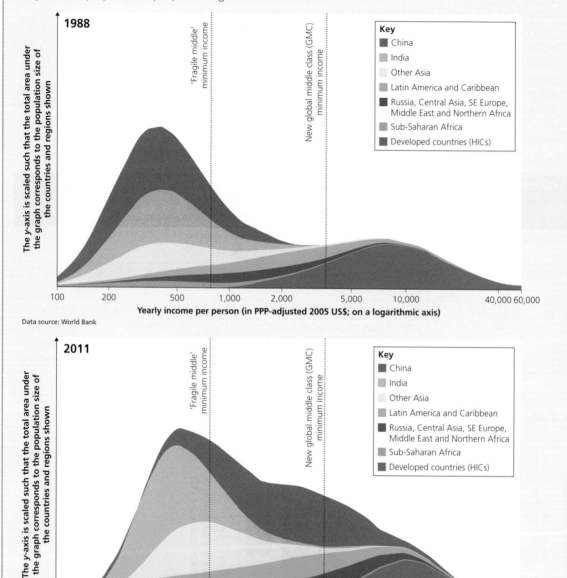

Data source: World Bank

▲ **Figure 4.12** Changes in global income distribution 1988–2011 show (i) a small increase in modal income and (ii) a growing proportion of middle-class earners

(a) Using Figure 4.12, analyse changes in (i) the modal income value and (ii) total world population size.

**GUIDANCE**

(i) The modal (most common) income increased slightly from around US$300 to just under US$500 (remember this is a logarithmic and not linear scale when making estimates). (ii) The total volume under the curve has increased significantly, indicating an overall increase in the total number of people alive (growth looks to have occurred in most of the developing-world regions and countries shown except China). Overall, total world population has seemingly grown by around 50 per cent.

(b) Describe the changes shown in the size and location of the global middle class (GMC).

**GUIDANCE**

The number and proportion of people with a middle-class income of US$3650 or greater has risen – from less than one-sixth to around one-quarter of the world's population by 2011, to judge by the data shown. There has been a significant locational shift. In 1988, the vast majority of the GMC lived in developed countries, with a small amount also in South America. But by 2011 around half of the GMC lived in emerging economies, notably China.

(c) The most recent data shown in Figure 4.12 are from 2011. Outline possible changes in the pattern of world incomes that may have occurred since then. Suggest reasons for the changes you have outlined.

**GUIDANCE**

There are many ways of answering this question – you are free to apply whatever relevant knowledge and understanding you choose to. Possible themes might include: further increases in the size of the GMC, not only in Asia but in North Africa too (this might be related to a growing volume of south–south trade and investment); some further shrinkage in the number of people below the poverty line globally (perhaps linked causally with progress towards global poverty-reduction targets); limited changes in income distribution in sub-Saharan Africa along with some increase in population size (this region still has high fertility rates).

# The wider dimensions of global development

Flows of money and ideas within global systems influence human development in many ways, not just economically. This section offers a brief look at some possible links between global systems and improving political freedoms in different local contexts.

### Global progress towards gender equality

When she was shot and injured by a Taliban gunman in 2012, Malala Yousafzai was actively campaigning for the right of girls to attend schools in the Swat Valley region of Pakistan (see Figure 4.13). A continuing campaign of violence against girls attending school has meant that many have been denied an education, despite it being a fundamental right for the citizens of Pakistan. The attack on Malala shocked Pakistan and the world – news travelled quickly through global media channels and social

▲ **Figure 4.13** Malala Yousafzai has become a figurehead for political development movements whose goal is improved gender equality

networks. As a result, Malala has since become a symbol of resistance against terrorism and the persistence of extreme social views that would deny women an equal right to education. Recently, efforts to improve the situation have taken place at varying scales.

- Local schools in Pakistan's Federal Administered Tribal Areas staged a day of action in 2014.
- Pakistan's government has promised to improve the number of girls participating in primary school.
- Pakistan is committed to the UN Sustainable Development Goals (see page 108), which include targets for education and gender equality.

### Affirmative action by TNCs

Some of the world's largest TNCs have taken affirmative action to try to protect the human rights of LGBT communities in some countries. A lack of equality for LGBT people can be viewed as an example of a development gap, which – along with a lack of equality for women or minority ethnic groups – occurs sometimes *within* particular countries.

Many TNCs are headquartered in world cities such as San Francisco and New York. They are creative places where people are more open to new ideas and diversity, according to an analysis carried out by the economic geographer Richard Florida. Some of these cities' technology and banking companies have led the way in trying to tackle prejudice in the workplace; a few have openly gay senior managers, including Apple's CEO, Tim Cook.

However, when US investment bank Goldman Sachs ran an LGBT recruiting and networking event at its Singapore office, a Singaporean government minister criticised the company for failing to 'respect local culture and context. They are entitled to decide and articulate human resource policies, but should not venture into public advocacy for causes that sow discord.' In Singapore – and many other countries in Asia, Africa and the Middle East – gay sex is still illegal.

For TNCs with a strong record of promoting diversity in the workplace in the countries they originate from, it can be a challenge to maintain a consistent global diversity policy. Campaigning too hard for equality could also affect a TNC's freedom to operate in some states; therefore affirmative action to tackle this development gap could become economically harmful. HSBC Bank has made a statement about this issue: 'We respect the law in countries in which we operate, but that doesn't prevent us having a global point of view. And our global point of view is to be very strongly, very firmly on the side of diversity and inclusion.'

Sometimes, governments and civil society organisations have taken affirmative action in support of LGBT rights too.

- In 2013, Uganda's government announced a new law making homosexuality punishable by death or life imprisonment.

- At the time, the UN High Commissioner for Human Rights said: 'It is extraordinary to find legislation like this being proposed more than 60 years after the creation of the Universal Declaration of Human Rights.'
- Shortly after, Uganda lost millions of dollars of international aid when donor countries cancelled payments in protest.
- The Ugandan government has since scrapped the death penalty, but homosexuality still warrants a prison sentence.

# ② Global migration and economic growth

▶ *In what ways can international migration help host and source regions to grow economically?*

One view is that international migration may help host and source countries alike to develop. Core–periphery theories in support of this argument were explored previously in Chapter 3 (see page 87). Table 4.3 shows the growth and developmental benefits of migration for both kinds of country.

| Host countries | Source countries |
|---|---|
| ■ Fills particular skills shortages (e.g. Indian doctors arriving in the UK in the 1950s).<br><br>■ Economic migrants willingly do labouring work that locals may be reluctant to (e.g. Polish workers on farms around Peterborough).<br><br>■ Working migrants spend their wages on rent, benefiting landlords, and they pay tax on legal earnings.<br><br>■ Some migrants are ambitious entrepreneurs who establish new businesses employing others (in 2013, 14 per cent of UK start-ups were migrant-owned). | ■ In Bangladesh, the value of remittances exceeds foreign investment. Unlike international aid and lending, remittances are a peer-to-peer financial flow: money travels more or less directly from one family member to another. This money flow helps the social development of communities who have previously been excluded financially from access to education and healthcare.<br><br>■ In time, migrants or their children may return, bringing new skills (some young British Asians have opened health clubs and restaurant chains in India, Bangladesh and other Asian countries). |

▲ **Table 4.3** Positive ways in which international migration promotes economic growth and development in host and source countries

The arrival of large numbers of low-skilled workers in a country can result in sizeable remittance flows directed towards the source country. Low-waged international migrants are drawn towards global hubs in large numbers. London, Los Angeles, Dubai and Riyadh are all home to large numbers of legal and illegal immigrants working for low pay in kitchens, construction sites or as domestic cleaners. Around US$500 billion of remittances are currently sent home by migrants annually. This is three times the value of overseas development aid.

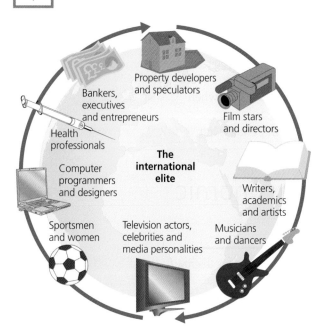

**▲ Figure 4.14** Global elite migration: these highly skilled migrants can play a vital role supporting the economic growth and development of host countries – but at what cost for source countries?

The movement of smaller numbers of high-skilled workers and high-wealth individuals has significant effects for both source and host countries alike. The variety of these elite migrants is shown in Figure 4.14. Their wealth derives from their profession or inherited assets. Some elite migrants live as 'global citizens' and have multiple homes in different countries. They encounter few obstacles when moving across borders. Most governments will welcome highly skilled and extremely wealthy migrants.

As Figure 4.14 shows, many elite migrants work in the knowledge economy, including writers, musicians and software designers. Skilled ICT professionals from the USA, India and elsewhere work in the UK's quaternary industry clusters in cities such as Bristol, London and Cambridge. This benefits the UK as a host nation for migration. India, however, is sometimes said to suffer from a so-called 'brain drain' effect: large numbers of its skilled medical and ICT workers have emigrated elsewhere.

## Migration and the development of global hubs

Demand for international migrant labour is concentrated often in particular global hubs located within host states. A global hub is a particularly important city when viewed *at both a national and global scale*. This is on account of the presence of the headquarters of major TNCs, globally

**▲ Figure 4.15** Oxford, UK, is a global hub for international migration. It attracts large numbers of overseas students and foreign-born academics. This has helped make Oxford the world's number one university city (in some academic rankings)

renowned universities, global financial and political institutions or other world-class assets. Global hubs like New York and Mumbai have gained in economic strength over time by attracting flows of foreign investment and the international workforce this brings. There are large numbers of foreign workers in London's Canary Wharf financial district. They play a vital role managing the European operations of American, Chinese, Indian, Japanese and Singaporean companies who have established offices there.

Some global hubs are megacities with over 10 million residents. Size is not a prerequisite for global influence, however. Smaller-sized global hubs that 'punch above their weight' in terms of their global reach include Washington, DC, and Qatar's Doha (which, as we have already seen, is a magnet for migration – see page 65). In

2018, Oxford University in the UK was named as the world's leading educational institution for the third year running (by *Times Higher Education* rankings): despite its small size, the city of Oxford is a globally influential place (see Figure 4.15).

## Migration flows and economic development along south–south corridors

Page 9 introduced the idea of south–south global system flows (movements

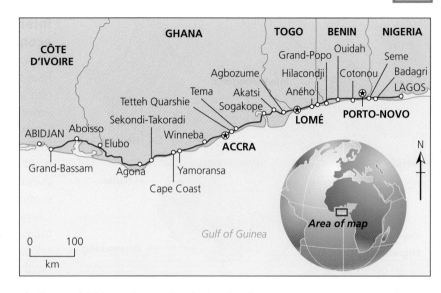

▲ **Figure 4.16** Large international migration flows are concentrated along the Abidjan–Lagos coastline (a major south–south migration corridor)

between different developing and/or emerging countries). UN data show that migration flows in south–south corridors are now equal to or greater in magnitude than south–north movements of people (see Table 4.4). Much of this consists of voluntary economic migration, although there are significant refugee flows also. Large movements between neighbour countries provide strong evidence for the way global flows are affected by the friction of distance (meaning that interactions are more likely between close neighbour states than distant countries – see page 82). Examples of south–south migrant flows include:

- Ghana to Nigeria (West Africa has very high levels of intra-regional international migration along a corridor stretching from Abidjan to Lagos – see Figure 4.16)
- Myanmar to Thailand (one of Asia's largest migration corridors).

In these and other cases, south–south migrants play an essential role in the continued growth of regional economic cores. In turn, remittances may contribute to the economic development of communities in migrant source regions.

| Migration corridor | Number of migrants (million) | Share of all international migrants (%) |
|---|---|---|
| South–south | 83 | 36 |
| South–north | 82 | 35 |
| North–north | 54 | 23 |
| South–north | 14 | 6 |

▲ **Table 4.4** Numbers of international migrants in the main global corridors, 2013

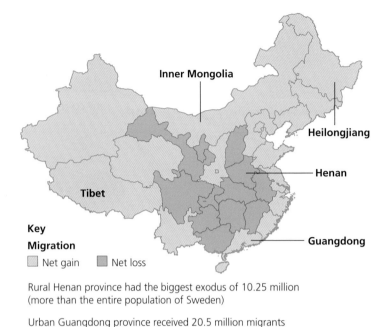

Inner Mongolia

Heilongjiang

Henan

Tibet

Guangdong

**Key**

**Migration**

☐ Net gain    ☐ Net loss

Rural Henan province had the biggest exodus of 10.25 million
(more than the entire population of Sweden)

Urban Guangdong province received 20.5 million migrants
(more than the entire population of Romania)

▲ **Figure 4.17** Rural–urban migration and the redistribution of China's
population, 1978–2010

## Internal migration and economic development

Alongside global migration flows, internal rural–urban movements play a vital role in a country's economic growth over time. For example, the mass migration shown in Figure 4.17 has undoubtedly been good news for China's economy overall. In 1978, on the eve of economic reforms, 20 per cent of China's population lived in cities; today, the figure is 55 per cent. The relocation of hundreds of millions of rural migrants to urban areas in turn attracted foreign investors who were keen to utilise this huge and modestly priced labour force. The Chinese government's authorisation of free movement can be viewed with hindsight as a rational economic decision that allowed the country to benefit from globalisation. The 'migrant miracle' supported three decades of rapid economic growth for China.

# ③ Communications technology and development

▶ *What contributions can ICT and data flows make to the human development process?*

Previously, Chapter 1 explored the rapid speed of arrival and maturation for new digital technologies (following Moore's Law – see page 36). Data flows play an increasingly important role in global development. Information and communications technology (ICT) has transformed every sphere of human life in both positive and negative ways. Pause and reflect for a moment on how recent advances in technology have affected people's participation in everything from retailing and education to hazard management and conflict.

## Different dimensions of growth and development

Nigeria had only 100,000 landline phones in 2001 (despite having a population of 140 million); by 2018 there were over 160 million mobile phone subscribers. Around 440 million people in African countries (about

44 per cent of the total) had mobile phone subscriptions in 2017 (although the global average remained higher, at 66 per cent, the gap is now much smaller than it used to be). Previously in this book we have encountered:

- migrants from poorer countries and regions transferring remittances home electronically and using phones to maintain their social networks (see page 10)
- the transformative power of mobile technology and M-Pesa banking in sub-Saharan Africa (see page 22).
- technically-led outsourcing and supply-chain growth on a global scale (see page 89).

This section briefly investigates further links between ICT and different strands of human development. Most of the supporting examples are also instances of technological leapfrogging.

## Data flows and economic development

New businesses have flourished in many global developmental contexts as a result of increased broadband speeds.

- The Indian city of Bengaluru (previously known as Bangalore) is a long-established technology hub, thanks to early investment in the 1980s by foreign TNCs such as Texas Instruments and domestic companies like Infosys (see Figure 4.18). Founded in 1981, Infosys had revenues of US$11 billion in 2018. It is one of the top 20 global companies for innovation, according to the US business analyst Forbes.
- More generally, global connectivity is a prerequisite for growth of the complex global production networks (involving offshoring and outsourcing) described in Chapters 1–3. Managers of distant offices and plants can keep in touch more easily (e.g. through video-conferencing). This has allowed TNCs to expand into new territories, either to make or sell their products. Each time the barcode of a Marks & Spencer food purchase is scanned in a UK store, an automatic adjustment is made to the size of the next order placed with suppliers in distant countries. This contributes to the economic growth of Kenya and other countries in the Marks & Spencer supply chain.
- China's impressive rate of economic development owes much to the way ICT networks have allowed the global supply chains that criss-cross its borders to operate smoothly and seamlessly.

Economic development processes are also supported by forms of microlending and other peer-to-peer linkages that would not always be possible without digital technologies:

- *Crowdfunding.* Small business start-ups in developing countries increasingly seek online pledges from the public to help get the funds they need. Individual investors from all over the world offer up-front support in return for future profits or products using services like Kickstarter; they can also sell goods and services globally using markets like eBay or Amazon.

 **KEY TERM**

**Leapfrogging** The World Bank defines leapfrogging as 'making a quick jump in economic development by utilising the latest technology'. The term describes what happens when a society moves straight to adopting a new, advanced form of technology without having invested in an earlier version (e.g. societies that adopt mobile phones without previously possessing landlines).

▲ **Figure 4.18** The Indian city of Bengaluru is widely portrayed in geographic literature as being a recipient of economic and other developmental benefits created by globalisation

- *Electronic remittances.* Somali migrants living in the UK send US$100 million to Somalia annually, relying on Barclays and other banks to transfer funds electronically.
- *Information transfers.* In Ghana, CocoaLink delivers advice and market price data to farmers using SMS text messaging.
- *Online disaster appeals.* After the Haiti earthquake in 2010, in the first 15 days the American Red Cross received US$29 million in the form of US$10 pledges sent through text messaging.

### Data flows and social development

ICT can help improve a developing society's access to health, education and energy supplies, especially in isolated rural regions.

- Remote healthcare is being provided in areas of the world where physical infrastructure is lacking; people in hard-to-reach parts of India can consult with a doctor using their mobile devices, for instance. Various start-up healthcare businesses and groups, including Nigeria's Ubenwa, are helping improve infant survival rates in rural regions of Africa and India. Logistimo is a Bengaluru-based company that provides health workers with an app which creates and uploads medical records for people in isolated areas. Khushi Baby is a non-profit organisation that provides pendants with an embedded microchip to children and mothers living in Udaipur, India. The pendants store each person's medical information. This allows visiting health workers to easily read a new patient's medical history.
- ICT is being used to improve so-called last mile (or longer) logistics in ways which help people in isolated and remote areas. Companies such as California-based Zipline are delivering medical supplies, including vaccines, blood and medicine, to villages in Rwanda by drone that are inaccessible by road. This can help fight malaria and measles (see Figure 4.19).
- Increasing numbers of people gain their education remotely by studying at a virtual school or university, or by enrolling in online MOOCs (massive open online courses).
- Charitable health and education institutions can raise funds online through direct appeals or as a result of individuals seeking sponsorship.
- Globally, over 70 million people in poorer countries were using mobile phones to purchase pay-as-you-go credits for 'off-grid' solar panels in 2016 (in Kenya, the service is provided by M-Kopa Solar). People have leapfrogged directly from firewood to renewable technology for their energy needs. They have skipped the need for grid-based electricity to be provided by the state.

### Data flows and political development

In some geographical contexts, such as autocracies, data flows play a role in the political development process. ICT serves as a tool to help campaigning for fundamental human rights and against state corruption.

**KEY TERMS**

**Last mile** In the UK, this is the last mile (or kilometre) that links any individual's home or office to a data network. This often gives services providers their greatest challenge because, in practical terms, it can be very hard to deliver fast broadband to some isolated places. In rural Africa, the challenges of connectivity may be far greater, with people living hundreds or thousands of kilometres away from the main network cables.

**Autocracy** A system of government where power lies mostly in the hands of a single individual or small group.

- The uprisings in Egypt and Tunisia during 2011 are sometimes described as 'the Facebook Revolution'. Movements against the regimes of Mubarak and Ben Ali, respectively, were organised online. The TV station Al Jazeera, based in Qatar (see page 66), also helped to publicise the protests. The downside of ICT for political development is the way fake news can undermine democracy too (see page 30).
- The political and ethical messages of non-governmental organisations (NGOs) find a global audience online, including campaigns by Amnesty International and Greenpeace. The global offices of large NGOs, and the research and journalism networks they depend on, are far more closely connected than in the past, thanks to ICT. Donations are easily collected online too.

The work and functioning of intergovernmental organisations (IGOs) is also enhanced by the ease with which information and publications can be disseminated. Websites for the EU, UN and World Bank contain a wealth of resources that aim to educate a global audience about issues ranging from climate change to international war crimes.

## A persisting digital divide

The existence of a global digital divide can stem primarily from the absence of people or a significant market in some places. Lack of demand has limited the roll-out of fibre optics to some isolated rural regions in developed, emerging and developing countries alike. In 2018, 100 million people were living in parts of African countries where there was no mobile phone service at all; and over 60 per cent of Africans (around 600 million people) lacked access to the internet. Although the global digital divide has lessened over time, more than 3 billion people remained excluded in 2018, including around half of Asia's population. Often, this takes the form of a rural–urban divide within individual countries.

- One main cause of digital exclusion is a lack of telecoms infrastructure. Neither fixed nor mobile broadband services may be available yet, especially in those rural areas where distance challenges are greatest.
- Poverty explains why many people remain excluded in places where services are now becoming available but must then be paid for.
- There are cases where global interactions are blocked purposely by the state (see page 49).

### KEY TERM

**Digital divide** The inequality of access to the internet that exists between different social groups in a country or between the citizens of different countries. A further distinction can be made between people who have mobile access to fast internet services and those who use mobiles only to make telephone calls.

◄ **Figure 4.19** Drones are a 'shrinking world' technology now being used to overcome 'last mile' developmental challenges such as healthcare access

## Limits to leapfrogging

Is ICT really the key to accelerated growth for developing communities? Can technology tackle the challenges of poor health, schools, energy supplies and transport infrastructure? Ban Ki-moon, the former UN secretary-general, has argued that that improved data flows and digital services mean that 'the next century now belongs to Africa'.

But is this view unduly optimistic? A persisting digital divide means there are limits to leapfrogging in some local contexts where phone and internet services remain unavailable or of poor quality. Indeed, new development gaps are sometimes opening up at a local scale between those isolated areas where digital technology is available and places where it is not.

Leapfrogging sceptics have also argued that the benefits of ICT are sometimes oversold to places where basic services are still unavailable. The danger in believing that new 'technological fixes' are just around the corner is that it excuses the persisting failure of governments to provide badly-needed essential infrastructure (see Figure 4.20). Many farms in sub-Saharan Africa lack irrigation or access to improved seed and fertilisers. Poor roads continue to hinder access in many rural areas. Although digital data transfers and drones are starting to help address some of the challenges of remoteness, why were adequate roads not built long ago?

Similar questions can be asked of leapfrogging in healthcare provision. Online medical advice is useful, to be sure, but would some communities not benefit more from being given the clean running water and hospitals they have been denied for way too long? Leapfrogging sceptics may therefore conclude that what developing communities would benefit from most is improved governance and increased capital expenditure on basic infrastructure, rather than new phone apps.

▲ **Figure 4.20** What should be the most important priority for isolated rural areas of sub-Saharan Africa: providing broadband access or building roads and hospitals?

Evaluating the issue

**▶ Evaluating the effects of global systems on geographic patterns of inequality**

## Identifying possible contexts

The focus of this chapter's plenary debate is inequality. What distinguishes geographical studies of inequality from work done in other disciplines is the guiding principle that social inequalities are also spatial (and vice versa). Geographers want to understand how and why spatial inequalities are produced and reproduced over time, and at varying scales. For example, important questions to ask when investigating a country's developmental and economic geography are:

- Do all local places within a country have the same level of income, wealth and opportunity?
- Do all people within a country or local place, including women, children and different ethnic groups, share the same economic and social opportunities?

In previous sections, we've observed a close relationship between globalisation and global development. The amount of new wealth created has exceeded population growth every year since the Second World War, suggesting there is more and more money to go around (see page 119, Figure 4.8). Many countries have been lifted out of poverty, judging by per capita GDP trends. However, the way new wealth gets distributed is often complex and uneven. New patterns of inequality shaped by global systems can be seen at varying scales and the distribution of 'haves and have-nots' does not necessarily correspond closely with the populations of different countries. *Increasingly, 'transnational' patterns of extreme wealth and poverty transcend national boundaries.*

Another important idea to focus on from the outset is the way *inequalities can deepen even as the overall economic outlook is improving.* Global development processes have helped make the world as a whole richer, while also lifting hundreds of millions out of poverty, but not all inequalities have lessened and, indeed, some have worsened. Think of it this way: if you raise a floor slightly while simultaneously lifting the ceiling by a far larger amount, what is the result? Both are now higher than they used to be, but the gap is greater than ever.

### Differing geographic scales and perspectives

There are numerous different manifestations of both decreasing and increasing inequality within contemporary global systems. This evaluation has two main geographic focus areas.

1 *Changes over time in average national incomes.* For example, many countries that were previously categorised as low-income have undergone significant economic development and are now reclassified as emerging economies (or 'middle-income countries' or 'economically developing countries'). This means that the income gap between these countries and many developed countries (sometimes called 'advanced countries' or 'high-income countries') has lessened.

2 *New patterns of extreme wealth and poverty that transcend national boundaries.* The wealthiest ten per cent of global income recipients received more than half of total global income in 2018, while the poorest ten per cent received less than one per cent. These two groups are made up of people living in every country. The former group is composed, typically, of well-connected individuals (from a global systems perspective). The latter group, which includes subsistence farmers in sub-Saharan Africa, may be far more weakly integrated into the global economy.

## The validity and reliability of inequality measures

Finally, before starting to review any evidence for inequality in detail, it is worth pausing to think critically about the validity and reliability of whatever data are used (see page 114). There are two separate challenges here. They are (i) deciding what constitutes inequality and (ii) working out how best to measure it.

**(i)** The first prompts an ontological (philosophical) question: what kinds of inequality (for example, income, wealth or happiness) should we be looking at, and why?

**(ii)** The second challenge is epistemological (methodological) and involves knowing how to collect and analyse data reliably. Possible indicators of economic and social inequality include per capita GDP, national HDI rankings, Gini coefficient scores, life expectancy data, gender index scores, etc. Any or all of this information could be collected. Data may then be manipulated using statistical techniques and measures such as mean, modal or median calculations.

Different conclusions may therefore be drawn about the extent of inequality between or within societies *according to which indicators are chosen and what tests and methods are used to analyse the data*. For example, poverty indicators for sub-Saharan Africa show how data can be manipulated to provide contrasting conclusions about inequality. The number of people living there in absolute poverty increased from 100 million to 200 million between 1981 and 2011. The region's total population also doubled from 200 million to 400 million. Using these data, three propositions can be generated about poverty in sub-Saharan Africa.

1. Poverty remained the same at 50 per cent.
2. The number of people living in poverty doubled.
3. The number of people living free of poverty doubled.

Each statement is true but there are varying implications about what has happened to inequality. This demonstrates how quantitative data can be manipulated in different ways to generate vying claims about whether inequality is growing or lessening over time.

# Evaluating inequality – a *national* perspective

Any geographical investigation of global development trends usually begins with an overview of the changing pattern of richer and poorer countries. One way of tracing changes over time in the global distribution of haves and have-nots is to observe changes in the classification of countries as high-income, middle-income or low-income. Figure 4.21 shows one development taxonomy proposed by the IMF in 2011. This threefold classification consists of advanced countries (ACs), emerging and developing countries (EDCs) and low-income developing countries (LIDCs).

This is a very different pattern from the unequal world of the 1970s shown in Figure 4.22. The Brandt Line was a widely-used analytical device which divided the world into a simpler 'haves and have-nots' pattern called 'the global north and south'.

- This book has shown just how far Asian countries like Indonesia (see page 60) and China (page 117) have come economically since the 1970s, along with other emerging countries. For example, the GDP per capita of the UK and China in 1980 was approximately US$28,000 and US$1,000 respectively (when adjusted for purchasing power parity, at today's prices). In 2018, the figures were US$45,000 and US$18,000. Instead of being nearly 30 times higher (as it was in 1980), the UK's per capita GDP is therefore now only two to three times higher than China's, suggesting a great lessening of inequality.

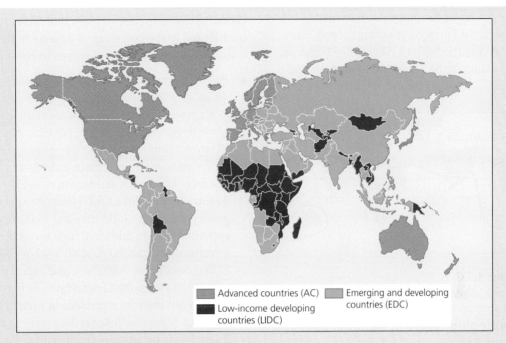

Advanced countries (AC)

Low-income developing
countries (LIDC)

Emerging and developing
countries (EDC)

**LIDCs** (low-income developing countries)
*Countries with lower average per capita incomes of around US$1,025 or below (2017). Agriculture plays a key role in their economies. Shallow integration into global systems, especially rural regions.*
Afghanistan, Armenia, Bangladesh, Benin, Bhutan, Bolivia, Burkina Faso, Burundi, Cambodia, Cameroon, Cape Verde, Central African Republic, Chad, Comoros, Democratic Republic of Congo, Republic of Congo, Côte d'Ivoire, Djibouti, Dominica, Eritrea, Ethiopia, The Gambia, Georgia, Grenada, Guinea, Guinea-Bissau, Guyana, Haiti, Honduras, Kenya, Kiribati, Kyrgyz, Republic Leo, People's Democratic Republic Lesotho, Liberia, Madagascar, Malawi, Maldives, Mali, Mauritania, Moldova, Mongolia, Mozambique, Myanmar, Nepal, Nicaragua, Niger, Nigeria, Papua New Guinea, Rwanda, Samoa, São Tomé and Principe, Senegal, Sierra Leone, Solomon Islands, Somalia, St Lucia, St Vincent and the Grenadines, Sudan, South Sudan, Tajikistan, Tanzania, Timor-Leste, Togo, Tonga, Uganda, Uzbekistan, Vanuatu, Vietnam, Yemen, Zambia
**EDCs** (emerging and developing countries)
*Currently experiencing higher rates of economic growth than in the past, usually on account of rapid industrialisation. Some EDCs correspond with the World Bank's 'middle-income' group. Others are high-income but newly-developed, including Qatar and Bahrain.*
Albania, Algeria, Angola, Antigua and Barbuda, Argentina, Azerbaijan, The Bahamas, Bahrain, Barbados, Belarus, Belize, Bosnia and Herzegovina, Botswana, Brazil, Brunei, Darussalam, Bulgaria, Chile, China, Colombia, Costa Rica, Croatia, Dominican Republic, Ecuador, Egypt, El Salvador, Equatorial Guinea, Estonia, Eswatini, Gabon, Guatemala, Hungary, India, Indonesia, Iran, Iraq, Jamaica, Jordan, Kazakhstan, Kuwait, Latvia, Lebanon, Libya, Lithuania, Macedonia, Malaysia, Marshall Islands, Mauritius, Mexico, Federated States of Micronesia, Montenegro, Morocco, Namibia, Oman, Pakistan, Palau, Panama, Paraguay, Peru, Philippines, Poland, Qatar, Romania, Russia, Saud, Arabia, Serbia, Seychelles, South Africa, Sri Lanka, St Kitts and Nevis, Suriname, Syrian Arab Republic, Thailand, Trinidad and Tobago, Tunisia, Turkey, Turkmenistan, Ukraine, United Arab Emirates, Uruguay, Venezuela, Zimbabwe
**ACs** (advanced countries)
*High-income developed countries. Office and retail work has overtaken factory employment, creating a post-industrial economy.*
Australia, Austria, Belgium, Canada, Cyprus, Czech Republic, Denmark, Finland, France, Germany, Greece, Iceland, Ireland, Israel, Italy, Japan, Republic of Korea (South Korea), Luxembourg, Malta, Netherlands, New Zealand, Norway, Portugal, Singapore, Slovak Republic, Slovenia, Spain, Sweden, Switzerland, United Kingdom, United States

▲ **Figure 4.21** Three groups of countries identified by IMF researchers and categorised according to per capita income and development dynamics. How far do you agree with this classification? For instance, should Qatar still be described as an EE rather than an AC? Some of the LIDCs shown, including Bangladesh, Kenya and Nigeria, are now categorised by other researchers as emerging economies. Views may therefore differ on the appropriateness of the IMF's analysis.

● Eastern European countries have also experienced periods of strong growth since the end of the Soviet Union in 1991 and the expansion of the EU in 2004 (when the A8 Group, including Poland, joined the EU).

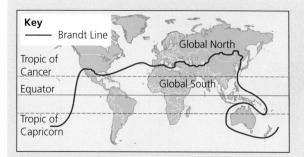

▲ **Figure 4.22** The Brandt Line: how we viewed global inequality in the 1970s

## The widening range of national average incomes

Although the income gap between developed and emerging countries has often lessened, some of the world's poorest countries have made little or no progress in recent decades. A few countries – particularly smaller states such as Qatar and Singapore – have experienced income growth in the region of 20,000 per cent since 1970. In 1965, Singapore's per capita GDP was the same as Gambia's, at just US$500 US dollars; in 2017, it exceeded US$90,000, while Gambia's was around US$1500 (all figures are PPP-adjusted). We therefore find evidence here of national-scale income differences actually *widening* over time.

In his analysis of global systems, the academic Manuel Castells argued that the poorest countries like Gambia are actually part of a so-called 'fourth world' of 'switched-off places' (when viewed from a networks perspective). They are states with weak levels of human capital, which may explain why they play only a very small role in global geographies of production and consumption (see Figure 4.23). Some of the world's very poorest countries have maintained a per capita GDP of US$500 or less for several decades.

Global systems operate in ways that have contributed to the obduracy of poverty in these countries. There are several factors to consider.

● *Poor access to markets within global systems.* The division of the world into trade blocs can help explain why national incomes remain low for some developing countries. For example, the EU protects its own farmers by placing import tariffs on food imports from outside its borders. As a result, farmers in non-EU countries like Kenya find it harder to get a good price for the food they sell to European supermarkets. In addition, high levels of government financial support enable many EU farmers to produce meat and vegetables cheaply. As a result, African farmers must offer to sell their own produce at even lower prices to firms like Tesco if they want to trade. There has been sustained criticism of the WTO (see page 52) for its ongoing failure to tackle EU and North American agricultural protectionism.

● *Low primary commodity prices.* According to the theory of comparative advantage (see page 83), a country's primary commodities (unprocessed food, timber, minerals and energy resources) should provide opportunities to trade with other countries, thereby generating the income needed for economic development to occur. In reality, this does not always happen, often due to over-production, and countries that trade only in agricultural produce and raw materials do not always gain a good income. This means they have insufficient money to import vital manufactured products from other countries. Longer-term development goals become even harder to achieve without computers for schools or specialised hospital equipment.

● *Out-migration of skilled, ambitious or talented people.* This may rob a country of its most valuable human capital, thereby creating even greater development challenges through the process of positive feedback (see page 37).

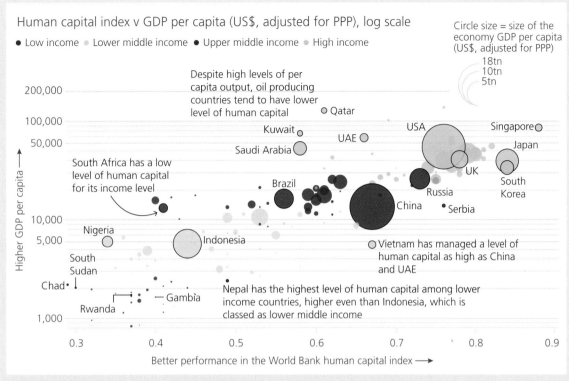

**▲ Figure 4.23** Marked global inequalities exist in average incomes and typical levels of human capital: there is also a strong correlation between these two variables. Note that the authors of this chart have used a range of well-chosen presentation methods to communicate their findings. What assessment could you offer of the chart's strengths? *FT Graphic; Source: World Bank. All data is 2014 or nearest*

- *Poor governance.* This has been a major contributing factor to the **under-development** of some poor countries. One view is that in the past, some developing countries lacked the human capital (e.g. skilled economists) needed to broker good trade deals (see page 159).
- *Unforeseen consequences of international aid.* Charitable donations have sometimes attracted criticism on the grounds that they do not solve poverty and can even perpetuate it instead. In the early 2000s, charitable donations of clothes to Zambia inadvertently devastated the country's own fledgling textiles industry. Sales of new clothes plummeted on account of well-intentioned gifts.

# Evaluating inequality – a *transnational* perspective

When we talk about the per capita GDP of this or that nation, it is all too easy to forget that it is actually a very crude average value which simply divides a country's total income by the number of people living there. The figure may, in fact, provide a poor picture of what life is *typically* like for most people. Indeed, few people may actually earn the mean average sum – in many countries, a relatively small elite group receive a very large share of national income, while the majority of people may have far less disposable income than the per capita GDP figure implies. We need to dig deeper into the data in order to really understand the extent of inequality both within and between countries.

## Asset and income growth for the global elite

One reason many commentators are concerned about globalisation's impact on inequality is the way the share of income going to the richest ten per cent of the world's people has risen far more rapidly than that of the poorest ten per cent. In other words, those who were already wealthy at the start of the 1990s have taken home a disproportionately large share of *new* economic growth in the years since then. This is because they were in the best position to make profitable investments using their existing capital: their purchases may have included property in world cities like London and Beijing (which tripled in value between 2005 and 2015) or shares in new technology companies (such as the FANGs – see page 37). In contrast, the majority of people's incomes have grown far more slowly over time because they have fewer financial assets to invest in wealth-making enterprises. Chapter 5 explores how housing and liveability injustices arise from this inequality of opportunity (see pages 154–57).

Figure 4.24 shows the share of national income taken by the top ten per cent of income recipients in G7 nations since 1981. Over time, the richest ten per cent have taken an increasingly large share of national income. The poor have not necessarily earned *less* overall as a result, but the rich have been able to multiply their earnings many times over, thereby pulling further away from the poor in terms of their luxurious lifestyle. Divisions in society have therefore widened.

### The role of migration

Migration can also play a role in increased income inequality in local contexts. Recently, the income gap between the very rich and very poor has grown wider in many world cities for two reasons.

- In-migration of large numbers of poorly-skilled workers produces a surplus of labour, which can depress wages further for low-

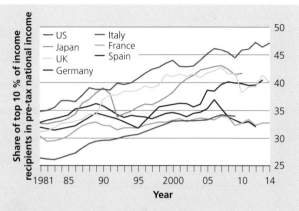

▲ **Figure 4.24** Rising inequality within G7 nations (years vary for most recent available data).
*FT Graphic; Source: World Inequality Database*

earners (employers can offer less money when more people are desperate for work).

- In-migration of highly-skilled workers can have the opposite effect: businesses begin to compete among themselves for the very best of the talent which is on offer, thereby pushing up wages at the top end of the spectrum (see Figure 4.25).

Migration flows may therefore become linked with rising inequality in cities like London or New York, unless there is state intervention (for example, minimum wage legislation).

### Putting it all together: the 'big picture' of global inequality (Lakner and Milanović)

A significant outcome of the patterns and trends described above is that the falling number of people living in extreme poverty globally has been accompanied by a rise in the number of people experiencing relative poverty in many societies. When the assets and earnings of the hyper-rich balloon in value, the average (per capita) level of income rises. As a result, some poorer people – whose earnings are static or have only risen modestly – are reclassified as having below-average incomes despite the fact they have experienced no material decline in the amount of money they make. To summarise, 'the rich get

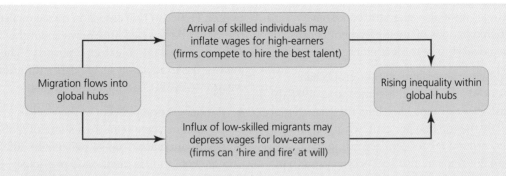

▲ **Figure 4.25** In some local contexts, international migration can lead to increased income inequality

richer while the very poorest do not – and so the gap widens' is perhaps the best summary view of contemporary geographies of inequality.

A fascinating view of this phenomenon was provided by Christoph Lakner and Branko Milanović in 2013. It is shown here in Figure 4.26. The 'elephant chart' (so named because of its slight resemblance to the large mammal) identifies two groups of people who have fared especially poorly during recent decades of global growth.

1   *The world's poorest ten per cent.* This consists mainly of people in low-income developing countries such as Chad, the Democratic

Republic of the Congo (DRC) and Eritrea where there has been limited foreign investment. Their incomes did not rise, or rose by just a few per cent, during the period shown of 1998–2008. This group – the world's lowest-earning ten per cent – may also include homeless people in UK cities (remember that this diagram shows *transnational* patterns of poverty and affluence).

2   *Those just below the top ten per cent.* These are citizens of richer developed countries who are most likely to define themselves as 'blue-collar', 'working-class' or 'ordinary people'. Their

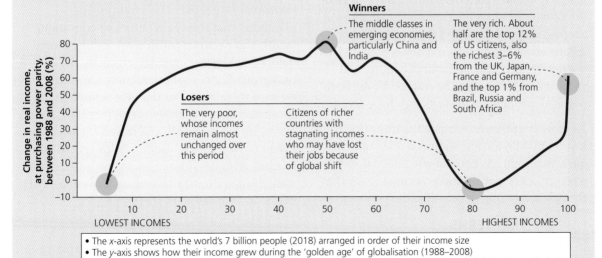

▲ **Figure 4.26** A different view of an unequal world: the Lakner and Milanović 'elephant chart' shows globalisation's winners and losers

incomes were also stagnant; some have suffered periods of unemployment because of the global shift of work to emerging economies.

According to these data, neither group has experienced a rise in real incomes since 1988. In contrast, other groups have benefited much more, especially the top one per cent, whose earnings grew by around 60 per cent according to Figure 4.26. You can also see income growth of 70–80 per cent during 1988–2008 for people in the middle of the global income range (i.e. the new middle class of emerging economies).

Not all of the changes shown are a direct effect of global systems of course. Lakner and Milanović argue that the slow growth of incomes for 'ordinary people' in advanced economies is explained by the combined effects of technological change, globalisation and the economic policies of governments. Moreover, it is hard to separate these three effects, as Chapters 1 and 2 showed.

## Reaching an evidenced conclusion

What can we finally conclude about the effects of global systems of patterns of inequality? To what extent has globalisation increased inequalities at varying scales or narrowed them by fostering greater economic growth and development? This is a notoriously contested area of enquiry and varying conclusions may be reached by different people depending on the facts they have access to.

As more countries have made the transition from low- to middle-income status, so one argument goes, the world has become less unequal. Yet some critics of globalisation continue to say that the rich are getting richer while the poor get poorer. How can this assertion be reconciled with the World Bank's headline news about the ongoing global fall in extreme poverty? The answer lies in an analysis of *relative* differences in wealth. In 2017, Oxfam calculated that:

- the richest one per cent of the world's population have seen their share of global wealth increase from 44 per cent in 2009 to 99 per cent (and, as mentioned previously, the richest eight billionaires possess the same wealth as the poorest half of humanity)
- nearly 1 billion people still live on less than US$1.90 per day.

By these measures, the world has never been less equal than it is today in some ways, and there has been an 'explosion in inequality'. But it is development gap *extremities* which have grown (the range of values between the world's very richest and poorest people and countries).

In closing, we should remember too that there is more to inequality than economics. Uneven access to opportunities, happiness and wellbeing remain global causes for concern and were the stimulus for the Sustainable Development Goals (see pages 101 and 108). In many places, people's life chances (the likelihood that they gain sufficient education, health and personal fulfilment) continue to be determined by their gender, ethnicity, social background or sexual orientation. Opportunities for men remain greater than for women in most countries; homosexuality is still illegal in many states, including Saudi Arabia and Uganda. The extent to which globalisation helps spread more tolerant attitudes, ideas and norms is therefore something we must not lose sight of either. This theme is returned to in Chapter 5.

 **KEY TERMS**

**Human capital** An attempt to quantify the net value of a population's skills. People's capabilities stem from the way their education, training, abilities and ideas are shaped by surrounding economic, social and cultural forces.

**Over-production** This occurs when too many countries grow the same crop. This over-supply pushes down prices globally. When crop yields are especially high due to good weather, the problem worsens. In some years, prices for coffee beans, cocoa beans or bananas have fallen very low, bringing misery to producer communities.

**Under-development** A theory that suggests some places are less developed than they might otherwise be on account of external interference such as colonialism and neo-colonialism.

**Liveability** An assessment of what the overall work–life balance in a place feels like, taking into account environmental, community, economic, housing and transport/commuting conditions.

**Relative poverty** When a person's income is too low to maintain the average standard of living in a particular society. Asset growth for very rich people can lead to more people being in relative poverty.

**Life chances** The opportunities each individual has to improve their quality of life. These opportunities are strongly affected by the way societies are managed and which polices governments adopt.

# Chapter summary

✔ Development encompasses a country's economic growth and also a wide spectrum of social and political changes. Views differ on the validity and reliability of different development indicators.

✔ Patterns of global trade and development correlate closely with one another. The development of emerging economies is linked with the growth of so-called south–south trade, in addition to north–south trade. China's recently accelerated economic development corresponds with the leading role it has assumed in global trade.

✔ There is strong evidence linking the growth of global systems with worldwide poverty reduction and the emergence of a new global middle class. Accompanying this, we see improved gender equality in many societies. However, not all places have benefited to the same extent from globalisation.

✔ International migration can help explain patterns of economic growth; both host and source countries can benefit from movements of people across borders. Migration is an important input into the economic systems of global hubs (world cities). South–south migration has played an important role in the growth of some emerging economies.

✔ The acceleration and diffusion of global data flows is linked with global development in varied ways. ICT leapfrogging provides local communities with new ways of raising capital and earning money; it can be used to improve social development (education and health) and mobilise democratic political movements. Not all societies have access to ICT, however, due to a persisting global digital divide.

✔ The effects of global systems on geographical patterns of inequality are complicated. Data can be interpreted in varying ways to produce contrasting arguments about whether inequality has increased or decreased. Inequality has often lessened between countries (according to per capita income measures). But there has been a marked rise in inequalities of income and wealth between the world's very richest and poorest people (irrespective of which countries they live in).

## Refresher questions

1. What is meant by the following geographical terms? Human development; south–south trade; extreme poverty; global middle class.

2. Explain how human development index (HDI) scores are arrived at.

3. Suggest reasons why some countries' per capita GDP data may be unreliable.

4. Outline reasons for global poverty reduction since the 1980s.

5. Describe changes over time in the distribution of the global middle class.

6. Outline ways in which different global players have helped promote equality for women and minority groups.

7. Using examples, explain how international migration can benefit (i) host countries, (ii) source countries and (iii) global hubs.

8. What is meant by the following geographical terms? Digital divide; leapfrogging; last mile.

9. Using examples, explain how online peer-to-peer lending can help tackle different kinds of development gap.

10. Explain why inequality can rise in a society even though average earnings are also rising.

## Discussion activities

1. The following question invites you to think synoptically (in other words, you can draw on all of the different human and physical geography topics you have studied). 'Global inequalities are lessening but local inequalities are growing.' Referring to evidence from both physical and human geography, how far do you agree with this statement?

2. 'Who wants to be a billionaire?' Conduct a whole-class debate based on the following facts.
   - Global systems operate in ways which have allowed a small group of billionaires to accumulate unimaginably large amounts of wealth.
   - Meanwhile, many pressing global challenges need tackling urgently (including climate change, persisting poverty in sub-Saharan Africa and rising international tensions).
   - Yet many governments say they cannot afford to deal with climate change and other urgent issues.
   - Should billionaires be obliged to tackle global challenges if they have the financial means to do so?
   - What different political, economic and moral viewpoints are there to consider?

3. In small groups, discuss the importance of scale when studying the geography of development. Important questions to think about include the following:
   - Do all local places within a country have the same level of development? If not, why not?
   - Do all groups of people share the same economic and social opportunities? If not, why not?

# FIELDWORK FOCUS

This chapter's focus on development and growth provides a conceptual foundation for fieldwork focusing on how financial flows originating in the UK may be helping the development process in other countries.

A *Surveying a diaspora community in order to explore the role of remittances in supporting the development process in another country.* If you study in a culturally-diverse school or college, it might be possible to build a small sample group of parents or grandparents (of other students you know) who originally migrated to the UK from other countries. Interviews could focus on whether remittances were ever transferred to family overseas, and how it helped them. There may be other forms of financial exchanges to think about too, such as the purchase of overseas properties or other assets. Be thoughtful, however, if you decide to do take this route: there are important ethical issues to consider. Questions will need to be phrased sensitively: you need to check your interviewees feel comfortable discussing money and other issues. Be clear upfront about what you want to talk about.

B *Investigating an important development issue, such as improving rights for women or LGBT communities in developing countries.* While these could be issues you are interested in studying, the fieldwork challenge will be generating *quality* primary data. Be careful to avoid producing an extended essay (based mainly on reading) at the expense of carrying out actual fieldwork. One approach might be to conduct interviews with a sample of the general public, perhaps focusing on whether people believe it is important for a country like the UK to promote and protect human rights in other countries. But would this generate data you can map or analyse with statistical tools? What other kinds of primary data could you collect?

# Further reading

Dorling, D. (2014) *Inequality and the 1%.* London: Verso.

Lakner, C. and Milanović, B. (2013) *Global Income Distribution: From the Fall of the Berlin Wall to the Great Recession.* World Bank Policy Research Working Paper 6719. Available at: https://openknowledge.worldbank.org/handle/10986/16935.

Pilling, D. (2018) The limits of leapfrogging. *Financial Times*, 13 August. Available at: https://www.ft.com/content/052b0a34-9b1b-11e8-9702-5946bae86e6d.

Milanović, B. (2011) *The Haves and the Have-Nots: A Short and Idiosyncratic History of Global Inequality.* New York: Basic Books.

Nielsen, L. (2011) *Classifications of Countries Based on their Level of Development: How it is Done and How it Could Be Done.* IMF Working Paper 11/31. Available at: www.imf.org/external/pubs/ft/wp/2011/wp1131.pdf.

Williams, G., Meth, P. and Willis, K. (2014) *Geographies of Developing Areas.* London: Routledge.

# CHAPTER 5

# Global injustice

Unequal flows of people, money, ideas and technology within global systems sometimes act to promote development but can also cause injustices and conflicts. This chapter:

- examines the adverse impacts of global investment and trade flows for different societies
- explores real and perceived injustices caused by international movements of people
- investigates environmental injustices suffered by different people and places
- evaluates efforts to tackle different kinds of local injustice associated with global systems.

## KEY CONCEPTS

**Social justice** Fair treatment of different people, measured in terms of wealth, opportunity and privilege. Social justice means that people's life chances (the likelihood that they gain sufficient education, health and personal fulfilment) are not unduly determined by their nationality, gender, ethnicity, background or sexual orientation.

**Injustice** When people, places or environments are treated in unfair ways.

**Risk** The possibility of a negative outcome resulting from a decision or process. For example, some groups of people are at heightened risk of exploitation, physical harm or homelessness because of the way global systems operate.

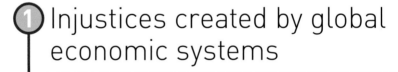

# ① Injustices created by global economic systems

▶ *What different kinds of injustice arise from global trade and investment?*

## Inequality or injustice?

The previous chapter focused on inequalities within global systems. For example, we know that already wealthy people are best positioned to invest money in global systems (for example, in property or companies), thereby gaining even greater dividends for themselves. As a result, inequality between the richest and poorest people keeps growing in many societies. But is this an example of injustice too? Many people would disagree. In most democratic societies, there is a broad consensus among citizens that it is fair for people to inherit family wealth (which can then be invested in new ventures). However, perspectives will differ on whether it is fair for people to inherit the *entirety* of large estates without paying more tax.

In any context, people are likely to disagree to some extent over what they view as fair treatment of themselves or others. Moreover, views about what constitutes injustice can change over time.

- In the past, unequal pay and voting rights for men and women were not always viewed as *injustices*. This was because of the prevailing view back then that biological differences justified unequal treatment. But because ideas about gender have since changed, unequal pay for men and women is now viewed by most people as a matter of injustice and not just inequality (see Figure 5.1).
- Similarly, there is now greater recognition that serious injustices were suffered in the past by indigenous people whose land was seized by outside forces. Between the 1500s and 1800s, colonial powers invaded large parts of Africa, Asia, Latin America, Australia and New Zealand. Many populations suffered occupation, displacement and even slavery. At the height of the British Empire, however, most British citizens would have viewed their country's overseas actions as perfectly just and fair, (many believed they were 'civilising' the world). Today, we may view the past more critically. For example, the 'Rhodes Must Fall' campaign is a political movement which raises awareness of historical injustices and aims to 'decolonise' universities around the world (see Figure 5.2).

**Inequality (but not injustice)** For example when one worker is paid more than another but has greater responsibilities or has worked longer hours

**Inequality *and* injustice** For example when a man and woman do identical work but for different wages

▲ **Figure 5.1** Some instances of inequality are also injustices

◀ **Figure 5.2** The 'Rhodes Must Fall' campaign began in South Africa in 2015 and soon spread globally. Students at the University of Oxford called for a statue of Cecil Rhodes to be removed from Oriel College. Rhodes used to be regarded as a great person who helped develop South Africa, but modern critics say he was guilty of imposing great injustices on African people

## Injustices arising from the financialisation of development

Recent work by academic geographers has drawn attention to injustices arising from the **financialisation** of global systems. The increasing influence of financial logic and players on the fabric of everyday life has

**Financialisation** A rising tendency to measure everything (including land, water and ecosystems) in terms of its economic value rather than by any other criteria. It is linked with the spread of neoliberal political values and the growing power and influence of businesses at all spatial scales.

▲ **Figure 5.3** A young girl helps her mother collect dirty water from an unsafe well outside Dar es Salaam, Tanzania. Financialisation of the development process has exposed vulnerable groups to new risks

**New international division of labour (NIDL)** The worldwide restructuring of productive activity which deepened after the 1980s. In essence, low-skilled and poorly-paid work migrated towards developing and emerging countries, while higher-paid skilled jobs (e.g. management and research) were retained or created in developed/advanced countries.

improved the wellbeing of some individuals and societies while simultaneously eroding the life chances of others. Chapter 2 examined how IGOs (the World Bank and IMF) have encouraged (and sometimes pressured) national governments to embrace neoliberal ideas and goals. One view is that this has created injustices because of the way vulnerable social groups become exposed to new risks as part of the financialisation process.

The unjust effects of financialisation are demonstrated by the cautionary tale of what happened to young girls from poor families when an attempt was made to improve water supplies in Tanzania during the early 2000s. Government-run services had fallen into disrepair in the 1990s but still managed to deliver safe water to some of the poorest households in capital city Dar es Salaam's slums. Clean water is essential if social development goals are to be met, such as improved school attendance for both boys and girls. Unsafe water results in illness and school absences. In the past before the water system existed, girls frequently missed school because they spent their days carrying buckets of water from wherever it could be found to their family home.

Tanzania approached the World Bank for help. As part of a deal, the World Bank insisted that Tanzania privatise its water services in return for a new US$143 million loan. Consequently, Dar es Salaam's water services were sold overseas to a British-led consortium called City Water which took over the day-to-day running of the city's water supplies. For the first time, water bills were issued to all households with access to drinking water. When some households could not pay their bills, they were disconnected. Due to the financialisation of their urban infrastructure, Dar es Salaam's poorest and most vulnerable families now reverted to the use of unsafe water sources and girls began missing school again (see Figure 5.3).

In 2005, the Tanzanian government successfully cancelled the contract with the consortium. Today, Dar es Salaam's water services are run locally once more but with support from several external players, including the African Development Bank. In 2012 the Indian government provided a US$178 million loan for water projects in Dar es Salaam. This is symptomatic of a wider shift among poor countries towards seeking support from new superpowers like China and India alongside, or in place of, the neoliberal Bretton Woods Institutions designed by the USA and its allies.

## Injustices arising from international divisions of labour

The global pattern of work has evolved well beyond the traditional worldwide division of labour (wherein farmers and miners of developing countries supplied raw materials for the factory workers of Europe and North America to process in value-added ways). As Chapters 1–3 showed, complex global networks of economic activity have grown over time. Global shift in the 1970s and 1980s created a **new international division of labour (NIDL)** based around both primary and secondary industries. Subsequently, geographical divisions of labour have also emerged for service sector and quaternary work, facilitated by advancement in ICT and data flows. Table 5.1 analyses the injustices arising from each industrial sector's spatial patterns and practices.

| Primary sector (agribusiness workers) | Commercial agriculture is a major global employment provider dominated by TNC giants such as Del Monte and Cargill. This industrial sector is no stranger to tough working conditions (see Figure 5.4). To judge by UK workplace and legal standards, injustices are widespread. Women and children are particularly vulnerable to exploitation in some contexts. Here are some examples:<br><br>■ In Thailand, commercial rice farmers work 12-hour shifts bent double in 38°C heat – yet some receive just US$2 a day for the repetitive work. At harvest time, Indonesia's prawn aquaculture labourers must work both day and night shifts on inhospitable tidal mud flats. In Costa Rica, banana crops are sprayed with pesticides while labourers are working in the fields.<br><br>■ Labourers commonly face workplace insecurity, never knowing in advance how many hours they will be needed for. The 'just-in-time' supply system used by UK supermarkets (see page 129) makes perfect financial sense for these businesses. But for women with daily childcare responsibilities (page 168), continued agricultural employment becomes an impossibility under irregular and unpredictable working conditions. This is a gendered manifestation of injustice.<br><br>■ According to Amnesty International, children as young as seven participate in small-scale cobalt mining in the Democratic Republic of the Congo (DRC). The materials they help produce end up in the batteries of Tesla and BMW cars. Most TNCs attempt to monitor their supply chains for unjust treatment of workers (see page 164) but far more needs to be done. |
| Secondary sector (factory workers) | Chapters 1 and 3 explored reasons why so much manufacturing work has been offshored or outsourced to countries where wages are relatively low and the law allows long hours to be worked, sometimes in conditions that would be deemed illegal in the UK.<br><br>■ Every year during the 1990s and early 2000s, an estimated 2500 metal-workers in Yongkang lost a limb or finger to the monotonous grind of machine work. This city gained a reputation as China's 'dismemberment capital' while assembling goods under contract for household names like Black & Decker (see page 150).<br><br>■ More recently, factories in Bangladesh and Vietnam have gained an unwanted reputation as places where unjust factory working practices are tolerated. In addition to health and safety risks, employment insecurity haunts workers with families to feed: temporary contracts are often the norm. Intensive production spurts occur in advance of Western holidays such as Christmas and Easter when more toys, clothes and other gifts are consumed. Following periods of high production, workers may find they are no longer required and are suddenly without an income. The consequences can be devastating for adults with families to feed.<br><br>■ That workers continue to tolerate such insecure and unjust working conditions in many countries is symptomatic of a fundamental lack in bargaining power with employers due to (i) a surplus of unemployed labour waiting in the wings to replace disobedient employees, (ii) low tolerance of trade unions by governments who fear **capital flight** by TNCs and (iii) the common practice of withholding wages owed in arrears in the event of an employee quitting. In one extreme case, female textile workers in Nicaragua testified they were sacked for joining unions; some claimed promotion could only be gained from management in exchange for sexual favours. |

| Tertiary sector (call centre workers) | Previous chapters explored how broadband internet has allowed offshoring of 'white-collar' (office-based) work. New opportunity has transformed lives in Bengaluru, India's fastest-growing city, with a global reputation for call centre telephone-enquiry work (see page 129). HP, IBM and American Express have call centres there, while large independent operators now conduct contract work for all kinds of firms, from travel companies to credit card providers. But work in this sector can be demanding.<br><br>■ Business is often conducted at night – due to time zone differences between India and customer locations in the USA or UK – sometimes in ten-hour shifts, six days a week.<br><br>■ Some employers demand workers adopt an artificial Westernised identity (such as an Anglicised name), hide their accent and conceal the call centre location from customers – all of which may cause psychological tension for staff. |
|---|---|
| Quaternary sector (research subjects) | Research and development of new information technology, biotechnology and medical science belongs to the quaternary sector of industry. The high skill level demanded by this fourth type of employment has not prevented its global shift, and in ways which are morally questionable.<br><br>■ Medical research is increasingly conducted overseas by large international pharmaceutical companies such as Pfizer and GlaxoSmithKline. African nations have often been the sites of clinical testing, including Cameroon, Guinea and Nigeria (as reported by Ben Goldacre in *Bad Pharma: How Drug Companies Mislead Doctors and Harm Patients*).<br><br>■ International welfare organisations are concerned that poorly-paid volunteers for drug trials in these countries are being exploited, serving as guinea pigs to test medicines they may never be able to afford. |

▲ **Table 5.1** Injustices for workers span all sectors of industry at a global scale. Practices are unjust insofar as workers in developing and emerging countries are treated in ways that would usually be viewed as morally and ethnically unacceptable in a developed/advanced country such as the UK

## 'Flying geese' (dynamic geographies of development and injustice)

**☞ KEY TERM**

**Capital flight** If workers in one country are awarded significantly improved pay and conditions, TNCs may choose to invest in countries where labour costs remain relatively cheaper.

▶ **Figure 5.4** Transplanting rice in Thailand: agricultural work involves hard labour and long hours for little pay

The phrase 'flying geese' conjures up a vision of birds flying in a stepped formation. Each group is followed by another. This evocative image – first used by Japanese economists – serves as a metaphor for changing patterns of world development. When the economies of one group of countries start to mature, capital flight occurs. This is because working conditions and pay improve as a state develops. Labour-intensive industries wanting low production costs move their operations to less-developed places.

As a result, geographies of injustice are constantly changing. In successive historical eras, different countries have become most strongly associated with unjust treatment of workers. The issues include excessively long working hours, 'dollar-a-day pay', DDD (dirty, dangerous and demeaning) occupations, slum housing (where diseases spread rapidly) and large gender pay gaps. Table 5.2 offers one view of this 'mobility of injustice'.

- Between the 1980s and early 2000s, factory employees in China suffered conditions similar to those seen in the UK during the nineteenth century. Chinese workers were making goods similar to those once produced in Europe and the USA but without the high levels of pay and workplace safety that European and North American workers now expect (making production there more expensive, partly explaining global shift).
- More recently, rising wages in China have prompted companies to relocate production to Bangladesh, Vietnam and other lower-cost Asian countries. Increasingly, this also includes Chinese-headquartered businesses investing overseas alongside Western TNCs (no longer a 'follower goose', China has now moved ahead in the 'flying geese' formation, like Japan and South Korea before it).
- This newest tier of emerging countries has become the focus of concern about workplace injustice. The 2013 collapse of the Rana Plaza building in Dhaka – resulting in the deaths of 1133 workers – was viewed as a 'wake-up call' about the urgent need for workplace reforms in Bangladesh (see page 151).
- If conditions improve significantly in Bangladesh, we may see capital flight towards western Africa in future years. Many economists believe countries like Kenya will be able to cash in their demographic dividend if global flows of FDI are redirected from Asia to Africa for lower labour costs.

 **KEY TERM**

**Demographic dividend**
A phase in the growth of a country's population that offers high potential for economic progress. A country's fertility rate falls as it develops economically. The result is fewer dependent children and relatively more productive teenagers and adults in the population. A large body of young, healthy and aspirational people can be a locomotive for economic growth.

| Era | Regions where unjust treatment of industrial workers was often widespread |
|---|---|
| 1800s–early 1900s | Western Europe and North America |
| Late 1940s | Japan |
| 1950s–1970s | First-tier newly industrialising countries (South Korea, Taiwan, Singapore and Hong Kong) |
| 1970s–1990s | Second-tier NICs (Malaysia, Thailand, Indonesia, Brazil, Tunisia and South Africa) |
| 1980s–early 2000s | China and Mexico |
| 2010s | India, Bangladesh, Vietnam, Pakistan and Ethiopia |

▲ **Table 5.2** One view of shifting geographies of development and workplace injustice

# ANALYSIS AND INTERPRETATION

Study Figure 5.5, which provides a view of the global shift of industrial injustice since 1844.

## Manchester, England, 1844

The work between the machinery gives rise to multitudes of accidents. The most common accident is the squeezing off of a single joint of a finger, somewhat less common the loss of the whole finger, half of a whole hand, an arm, etc., in the machinery. Lockjaw very often follows, even upon the lesser among these injuries, and brings death with it. Besides the deformed persons, a great number of maimed ones may be seen going about in Manchester; this one has lost an arm or a part of one, that one a foot, the third half a leg; it is like living in the midst of an army just returned from a campaign. In the year 1842, the Manchester Infirmary treated 962 cases of wounds and mutilations caused by machinery, while the number of all other accidents within the district of the hospital was 2,426. What becomes of the operative afterwards, in case he cannot work, is no concern of the employer.

Frederick Engels, *The Condition of the Working Class in England in 1844*

## Yongkang, China, 2003

Yongkang, in prosperous Zhejiang Province just south of Shanghai, is the hardware capital of China. People all over the world see Yongkang-made parts. They are the nuts and bolts of hundreds of brand name products, like Bosch, Black & Decker, and Hitachi. Yongkang, which means 'eternal health' in Chinese, is also the dismemberment capital of China. Unofficial estimates run as high as 2,500 accidents here each year.

Some factories resemble operations at the dawn of the industrial age, where migrant labourers use rudimentary machines that can sever the limbs of those who succumb to momentary distractions.

Legally, the loss of all fingers on one hand is a sixth-degree injury, mandating compensation of 200,000 yuan, or about $24,000. In practice, most owners reach settlements with their employees for nominal amounts and pay their bus fare out of town.

'China's workers risk limbs in export drive', *New York Times*, 7 April 2003

▲ **Figure 5.5** The global shift of industrial injustice

(a) Using Figure 5.5, compare the two case studies of industrial injustice.

### GUIDANCE

This question is targeted at A-level assessment objective 3 (AO3), which measures students' ability to work with information and data. The two texts provide opportunities for you to manipulate and compare quantitative data (look for similarities in the number of injuries, for example). You can also compare the words or phrases used, which provide qualitative evidence for the severity of the injuries and injustices suffered. Because the question is focused on injustice, you should attempt to exemplify this concept using as many different supporting ideas as possible.

(b) Explain why Chinese workers suffered similar injustices in 2003 to those experienced by English workers in 1844.

### GUIDANCE

This question provides an opportunity for you to apply your knowledge and understanding of the global shift of labour-intensive factory work. Important themes include rising labour costs and heightened safety standards in developed-world factories. In turn, injustices for workers have also undergone a global shift, as shown. You might additionally mention that the two case studies shown form part of a bigger 'flying geese' picture of global development (see page 148).

# CONTEMPORARY CASE STUDY: TACKLING INJUSTICE IN BANGLADESH'S TEXTILE SECTOR

The collapse of the Rana Plaza building in Dhaka, Bangladesh in 2013 led to the deaths of 1133 textile workers (see Figure 5.6).

- On the day of the collapse, workers were sent back into the building by Rana Plaza's managers to complete international orders in time for delivery, even though major cracks had appeared overnight in the building.

- Walmart, Matalan and other major TNCs regularly outsourced clothing orders to Rana Plaza.

Since then, many European TNCs have signed the Accord on Fire and Building Safety in Bangladesh, which is a legally binding agreement on worker safety. These companies now promise to ensure safety checks are carried out regularly in all Bangladeshi factories that supply them with clothes.

- The Accord states that signatory TNCs are 'committed to the goal of a safe and sustainable Bangladeshi Ready-Made Garment (RMG) industry in which no worker needs to fear fires, building collapses, or other accidents that could be prevented with reasonable health and safety measures'.

- The agreement covers all suppliers producing products for the TNCs. It requires these suppliers to accept inspections and implement remediation measures in their factories if unacceptable safety risks are discovered.

Many TNCs that outsource work to Bangladesh have not signed up yet, however, despite the fact that the Accord will result in as little as US$0.02 being added to the production cost of a T-shirt. Profit maximisation dominates the decision-making of the non-signatory companies. In contrast, the UK's H&M clothing company not only signed the Accord but has additionally campaigned for Bangladesh's government to strengthen safety-at-work legislation and to increase the country's minimum wage.

▲ **Figure 5.6** The collapse of the Rana Plaza clothing factory in Bangladesh in 2013. Rana Plaza had an outsourcing relationship with many major TNCs

# ② Injustices arising from movements of people

▶ *What kinds of injustice arise for people and places because of international migration?*

## International migration and injustice

Both migration and tourism can have adverse impacts for people and places. Perspectives usually differ on whether these movements of people are, on balance, more positive than negative. The manager of a small business who wants to recruit cheap labour may have more positive feelings towards migration than an unemployed person who believes (rightly or wrongly) that migrants have caused local job or housing shortages.

Migrants may feel they suffer unjust treatment too. All around the world, migrant labourers carry out work local people would prefer not to do. In return, they may become poorly-paid victims of xenophobia. You can review previous sections of this book for evidence of possible injustices arising from migration for individuals and societies. For example, is the 'brain drain' effect (see page 126) a form of injustice for developing and emerging countries? Or do remittances offer fair compensation for what may be a permanent loss of a nation's talent?

### Global refugee movements

Many millions of people worldwide now live in refugee camps. Figure 5.7 shows the distribution of Syrian refugees in Europe and the Middle East in 2016. Those who were forced to flee their homes and possessions have clearly suffered a great injustice (see pages 58–59). Equally, citizens of Turkey and other states with large numbers of refugees may view the pressure put on their own country as an injustice too (see Table 5.3).

| Injustices for refugees | ■ In many refugee camps, adults cannot work as there are no opportunities to make a living. |
| | ■ The Universal Declaration of Human Rights (UDHR) says that everyone has a right to education. Yet refugee children may cease to be schooled, with highly damaging long-term impacts for these individuals. By one estimate, half of all refugees are aged under 17 and as many as 90 per cent no longer receive any education or a satisfactory schooling. |
| Injustices for host countries | ■ The majority of refugees do not attempt an ambitious long journey to a distant country. They instead travel no further than the nearest state neighbouring their home country. Figure 5.8 shows the disproportionate burden shouldered by Turkey, Lebanon and Jordan resulting from the Syrian refugee crisis. Many residents of these countries say it is unfair that they must support so many refugees. |
| | ■ Although all EU states are obliged to take in refugees because they have signed the UDHR, some European governments have done far more than others to help. Germany accepted over 600,000 Syrian refugees between 2010 and 2016. In contrast, some EU states have sheltered merely hundreds. Many German citizens regard this as an injustice: they believe the weight should have been shared far more evenly. |

▲ **Table 5.3** Different kinds of injustice linked with global refugee movements

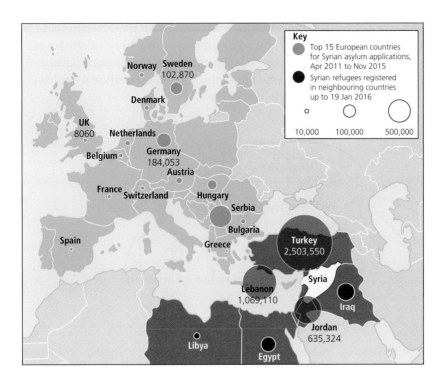

◀ **Figure 5.7** The distribution of Syrian refugees in Europe and the Middle East, 2016. Is it unfair if some countries provide shelter for large numbers of refugees while others do little to help?

## Migration and modern slavery

Flows of people in global systems give rise to injustices of varying severity. Among the worst are cases of human trafficking and modern slavery. Human trafficking can be seen as 'one of the dark sides of globalisation' according to the Institute of Labor Economics (IZA) in Bonn, Germany. The US Department of State has estimated there are more than 12 million victims of human trafficking worldwide, while crime agency Interpol categorises human trafficking as the third-largest transnational crime flow following drug and arms trafficking (see Figure 5.8).

Female victims of trafficking found in the USA come from 66 countries (China, Mexico and Nigeria, for example). Most trafficked women are forced into commercial sexual services, which is a form of modern slavery.

Although data flows and ICT help promote global development (see Chapter 4), they have sometimes played a role in modern slavery, too. Police forces in many countries are concerned by the rising use of Facebook and other social media for fraudulent and exploitative online recruitment of workers who risk becoming victims of modern slavery. There are recorded cases of modern slavery involving people from Romania and Nepal who were duped by online advertisements offering unrealistically high pay and were subsequently trafficked to Sicily and Kuwait respectively.

Modern slavery is believed to be rife in UK offshore waters. Foreign workers who do not have permission to live inside the UK are legally allowed to work on boats at distances where they are beyond the reach of checks by

 **KEY TERMS**

**Human trafficking** The recruitment and transporting of people into a situation of exploitation through the use of violence, deception or coercion. It is a highly gendered form of crime because victims are overwhelmingly female.

**Modern slavery** When a person is under the control of another person, who applies violence, financial penalties or other means to maintain the exploitation. Migrants are especially vulnerable to the risk of modern slavery when they cannot speak the local language and so cannot explain their situation and get help.

the police or UK welfare officers. This makes it hard to protect their human rights and to monitor what, if any, payment they receive. The treatment of some fishing-fleet workers from West Africa amounts to modern slavery (there are reports of workers who are victims of human trafficking or have had their passport stolen by their employer).

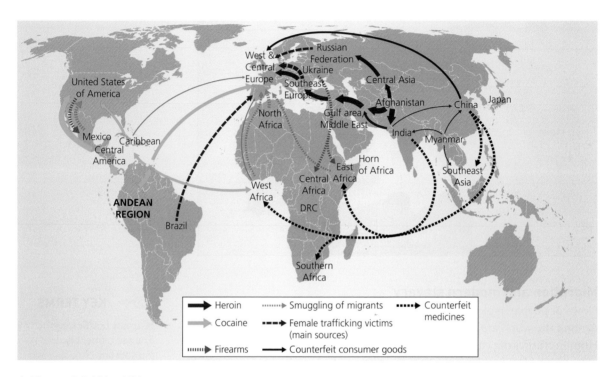

**▲ Figure 5.8** Using UN data, this map shows illegal flows of goods and people, including human trafficking. Each arrow width is proportional to the estimated market value. What kinds of injustice might be (i) causes and (ii) consequences of these flows?

## Housing injustices in global hub cities

Chapter 4 explored the idea of global hub cities (see page 126). Since the global financial crisis (GFC), prices have soared upwards in the world's most highly valued property markets. Prices rose on average by 30–50 per cent in London and other global hubs between 2010 and 2017. Shanghai recorded a staggering 150 per cent increase.

One result is that ordinary citizens have been left behind. Moreover, inequalities linked with the housing market have an important generational dimension. Homes are now so costly that Millennials (those born between 1981 and 2000) are often failing to get a foot on the property ladder: house prices may now be permanently beyond their reach. In contrast, older people bought their homes when costs were relatively much lower than today. Over time, they have watched their property assets multiply in value. Some people may view this outcome as a cause of inter-generational injustice.

Figure 5.9 shows the causes of rising housing costs within global systems. Hub cities act as magnets for low-skilled and high-skilled migrants alike.

Additionally, flows of money from property investors have helped create a positive feedback loop that has driven prices even higher.

- Many high-wealth individuals (in numerous countries) lost money during the GFC.
- In the aftermath of the GFC, they looked for safer investments (with their remaining capital).
- Investing in property in global hubs was seen as a low-risk and potentially high-return strategy.
- As a result, the property prices in most hub cities have soared upwards in a synchronised way.

In summary, global flows of people and money have driven competition for housing in global hubs to new heights, resulting in property prices which are now well beyond the reach of ordinary, younger people (see Figure 5.10).

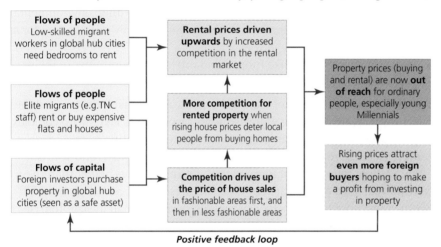

◄ **Figure 5.9** Global flows have led housing markets to 'overheat' in global hubs like London, Shanghai and Sydney. Property prices have risen beyond the reach of many younger people and lower-income groups

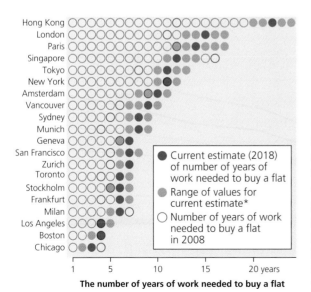

*The number of years of work needed to buy a flat*

*\*Uncertainty range due to differing data quality*

◄ **Figure 5.10** Changes in the number of years a skilled service worker needs to work to be able to buy a small 60m² flat near the centre of a global hub city. Wages and prices for flats have both grown since 2008, but the ratio has not stayed the same. It will now take longer for younger workers to save enough money to buy a flat

## Over-tourism and the injustice it creates

The global tourist industry generates huge international flows of people each year. Accompanying flows of money help transfer wealth between societies and may additionally encourage positive social and cultural changes in some contexts. Tourism also brings new pressures and injustices, though. Increasingly, the word 'over-tourism' is being used to describe the way excessive visitor growth has injurious effects for local (non-tourist) societies. Residents bear the cost of tourism growth, especially in places where ill-prepared planning authorities have failed to recognise that visitor limits are needed or lack the power to impose them. As a result:

- local communities are priced out of the property market where they live or have less disposable income (due to excessive housing and living costs)
- people's wellbeing is reduced because of the declining liveability of their home place; symptoms include year-round overcrowding, environmental damage and infrastructure strain (things tourists only need endure for a few days or weeks)
- souvenir shops spread like fungus and rob local communities of their own 'sense of place'.

The global and local flows of money that drive the tourist industry (and the injustices it creates) include (i) investments by powerful TNCs with complex tourist supply chains and (ii) property speculators profiting from **Airbnb rentals** (see Figure 5.11). According to the United Nations World Tourism Organization, international tourist numbers are 40 times higher than they were 60 years ago and tourism now contributes ten per cent of global GDP. Yet the most popular places to visit remain eternally small in size. Tourists converge, as they always have done, on key landmarks and historical districts – often with narrow pre-industrial streets – but in ever-increasing numbers and densities (see Table 5.4).

**KEY TERM**

**Airbnb rentals** A peer-to-peer economic service where individual homeowners let out rooms or whole properties directly to tourists online.

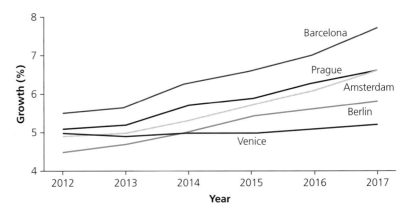

▲ **Figure 5.11** Estimated Airbnb overnight stays in popular EU cities, 2012–17. Entrepreneurs are buying up houses and flats to rent to tourists, driving property prices beyond the reach of ordinary local residents

Injustices arising from over-tourism are likely to increase not lessen in future years because:

- rising affluence is opening up popular destinations to new middle-class visitors from India, China, Brazil and other emerging economies
- the transport systems that enable international visitor flows have become truly global in reach and impact
- the internet amplifies the popularity of already-popular places; 'must-see' reputations can grow exponentially via feedback loops (social media posts about places result in more visitors who – in turn – add even more posts).

| Venice | 60,000 daily visitors have made the small city an unliveable place for many residents. The city may lose its coveted Unesco World Heritage Site status unless it acts soon. |
|---|---|
| Thailand | Andaman Sea beaches have suffered cultural and environmental erosion, including harm to coral reefs. Maya Bay alone receives 4000 daily visitors. |
| Iceland | Between 2010 and 2017, the number of arrivals quintupled in size to reach 2.5 million. But Iceland's own population is just 330,000. Housing costs have soared for locals in the small capital city, Reykjavík. |
| Barcelona | A record 30 million international tourists in 2016 left many of Barcelona's 1.6 million residents feeling overwhelmed by out-of-control visitor flows (see Figure 5.12). The injustices they suffer include: declining purchasing power parity for services and housing (in comparison with external speculators and visitors); a lost sense of belonging; and the declining liveability of their city. |

▲ **Table 5.4** Issues for over-tourism hotspots include concerns about social and environmental justice

▲ **Figure 5.12** Barcelona suffers from tourists congregating at high densities in historic districts such as El Raval and outside the Sagrada Família, the large church designed by Catalan architect Antoni Gaudí

 # Environmental injustices

▶ *What different kinds of environmental injustice do some societies suffer because of the way global systems work?*

Many different environmental injustices occur within global systems.

- Some environmental injustices derive *directly* from the actions of major global players (including TNCs, powerful states and wealth funds) and the negative effects of global flows of money, people and materials on the physical environment of particular places. This category includes the unjust impacts of land grabs, **extractivism** and transnational waste movements.
- Other environmental injustices derive *indirectly* from the operation of global systems. Foremost among these are climate change impacts on vulnerable societies whose limited use of technology and transport

 **KEY TERM**

**Extractivism** A 'mode of accumulation' that dates back at least 500 years. Natural resources are removed and transported elsewhere before being used. This means that the value added by processing and manufacturing is denied to the people and places where the raw materials were extracted.

means they have a negligible carbon footprint yet may bear the brunt of global warming impacts.

## Land grabs and extractivism

In the early 2000s, the Chinese government embarked on a programme of land acquisition in poorer countries, including Cuba and Kazakhstan. They were not alone; others, including South Korea and Saudi Arabia, started similar ventures in response to heightened concerns about long-term national food security.

▲ **Figure 5.13** A representation of land grabbing in Kenya

- For example, the Saudi Star company spent US$200 million acquiring and developing an area of land equivalent to 20,000 sports pitches in Ethiopia during the early 2010s (in total, Ethiopia's government has leased 2.5 million hectares of land). The land, in a region named Gambella, is now used to grow wheat, rice, vegetables and flowers for the Saudi Arabian market. Some Ethiopian farmers belonging to the Anuak ethnic group were required to relocate elsewhere and not paid any compensation, even though their families had farmed the land for centuries.
- Similarly, some indigenous communities in Kenya, such as the Ogiek people who have lived in the Mau Forest for generations, have struggled to gain recognition of their rights (see Figure 5.13).

Land grabs and extractivism are linked with injustice, but exactly how 'fair' or 'unfair' are these practices? Indigenous subsistence communities may have no officially recognised legal claim to their ancestral land. Often, they lack the literacy and education needed to assert and defend their rights in a court of law. Around the world there are many instances of unjust land grabs resulting in social displacements and refugee flows. You may be familiar already with the example of Amazonian rainforest tribes losing their land to logging companies.

▲ **Figure 5.14** Large areas of the USA are contested landscapes. Land that once belonged to Native American tribes passed into the hands of European settlers, who used newly-written property laws to justify their land grabs. Today, the legal position of Native Americans remains unclear when it comes to some issues such as oil and gas rights

A land grab may be perfectly legal, but that does not make it right (see Figure 5.14). It is well beyond the scope of this book to explore all of the philosophical and practical arguments which relate to this topic. However, some interesting questions to reflect on are included in this chapter's discussion activities (see page 171).

## Extractivism and environmental injustice in equatorial Africa

Georges Nzongola-Ntalaja (Professor of African Studies, University of North Carolina) has written that:

*Colonialism established a system of mineral exploitation that consisted of extracting raw materials for export, with little or no productive investment in the country from which they were extracted, and little or no effort to protect the environment. This system has remained intact since independence as a national curse, in that the country's enormous wealth attracts numerous outsiders who eventually find local collaborators to help them loot the country's natural resources.*

This quotation includes two important understandings about environmental injustice.

First, some countries have suffered a persisting injustice called the resource curse. In central and western Africa, natural resources have often been associated with conflict and injustice instead of growth and development.

**KEY TERM**

**Resource curse** The view that natural resource endowment may retard rather than accelerate economic and social development for some countries or regions, on account of the role resources often play in triggering war, corruption or the neglect of other development paths.

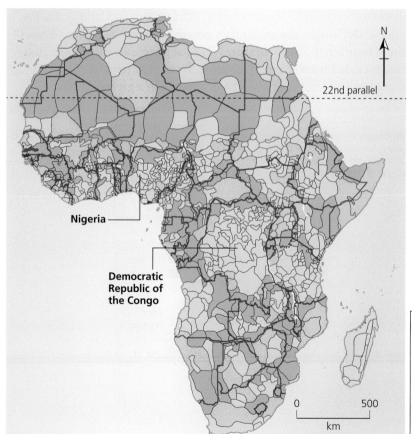

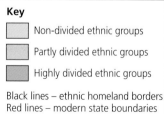

◀ **Figure 5.15** The territories occupied by different ethnic groups in Africa that preceded colonialism and the creation of modern African nation states. The extent to which individual ethnic groups or state governments own and control the land and natural resources in each territory raises difficult-to-answer legal, economic and ethical questions

Nigeria

Democratic Republic of the Congo

22nd parallel

N

0    500
km

**Key**

☐ Non-divided ethnic groups

☐ Partly divided ethnic groups

☐ Highly divided ethnic groups

Black lines – ethnic homeland borders
Red lines – modern state boundaries

- For example, the Democratic Republic of the Congo (DRC) – a country that was savagely colonised by Belgium in the late 1800s – is rich in natural resources that include copper, cobalt and diamonds as well as rare earths including coltan and niobium.
- However, it remains one of the least politically stable and ethnically fragmented regions in the world (see Figure 5.15), with 5 million lives lost to conflict and invasion between 1998 and 2007. Neighbouring countries' armies, including those of Uganda and Rwanda, repeatedly entered the DRC, ostensibly in support of either government forces or rebel groups. Once on DRC soil, however, these armies sometimes seized resources for themselves. Militia groups forced farmers and their families to leave their land if diamonds or metals were thought to lie beneath – or, worse still, pressed them into forced labour as miners.
- In 2017, the DRC was ranked 176th in the human development index. Life expectancy was just 59 and per capita GDP a mere US$444, with most people living on less than US$1.90 a day. Given this is one of the world's most resource-rich countries, such poor developmental outcomes surely represent a grave injustice.

The second important point made by Nzongola-Ntalaja is that 'local collaborators' have often played a role in extractivism and land grabs. For example, a TNC might pay a country's government for the right to extract resources from a populated rural area, but the people who live there find they are excluded from the profit-sharing. From their point of view, the collaborating TNC and politicians have done them a great injustice.

▶ **Figure 5.16** The Niger Delta: global oil companies and Nigeria's government have profited from this area's oil resources. But the indigenous Ogoni people have had to fight extremely hard to gain a share of the wealth

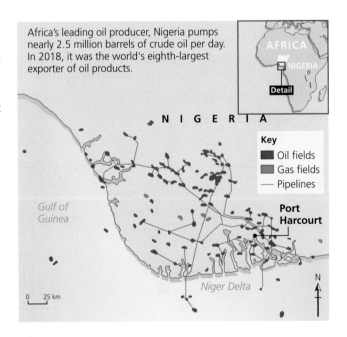

Africa's leading oil producer, Nigeria pumps nearly 2.5 million barrels of crude oil per day. In 2018, it was the world's eighth-largest exporter of oil products.

AFRICA

NIGERIA

Detail

**N I G E R I A**

**Key**
- ■ Oil fields
- ■ Gas fields
- — Pipelines

*Gulf of Guinea*

**Port Harcourt**

*Niger Delta*

N

0    25 km

- Nigeria's Niger Delta oil fields are a highly polluted site. Around 7000 oil spills occurred there during the 1980s and 1990s as a result of poorly-maintained pipelines owned by foreign companies (see Figure 5.16). This brought ruin to Ogoni people's farming lands. Indigenous writer Ken Saro-Wiwa led the protests that gained media attention; he was executed by Nigeria's government in 1995, causing an international outcry.
- Western oil firms, including Royal Dutch Shell and ExxonMobil, work the oilfields; in return, the Nigerian government receives around US$10 billion annually in revenues. In contrast, the Ogoni had gained little and lost much until recently. Following tireless campaigning (and helped by Amnesty International), in 2015 the Ogoni people finally gained some compensation from the Nigerian government and large oil companies for the damage done to their land. Royal Dutch Shell agreed to pay US$70 million in compensation to 15,600 farmers and fishermen whose lives were devastated by two large oil spills. However, the Ogoni argue this is only a fraction of what they justly deserve.

## Transnational waste movements

Many discarded and broken consumer items originating in high-income countries are transported to lower-income countries for disposal and recycling. For example, the vast majority of global ship-breaking (the process of dismantling an obsolete naval vessel) takes place in Bangladesh and India, where there are cheaper labour costs and fewer health and safety regulations. This trade has economic benefits for waste reprocessing companies in recipient countries. However, there may be significant health costs for workers in **pollution havens**.

Nowhere is this more apparent than in the **e-waste** sector – one of today's fastest-growing transnational waste streams. Old items can be dismantled and the waste burnt to extract valuable metals, including gold, silver, chromium, zinc, lead, tin and copper. Some e-waste is processed safely in formal-sector purpose-built facilities (for example, much of the UK's e-waste is exported to Belgium and Poland). However, large volumes of e-waste from high-income countries are processed by poorly regulated informal-sector players in developing and emerging countries. There, the work creates health hazards for those who undertake it and critics of global e-waste trade have highlighted the injustices it brings to some people and places.

Large numbers of people in India and China, including children working in family-run workshops, take part in informal e-waste recovery. Some of the worst and most widely reported problems have arisen in Ghana (see page 162).

 **KEY TERMS**

**Pollution haven** A low-cost location where global waste flows terminate. Movement patterns are explained by the relative cost of complying with environmental regulations in different places, for example in relation to waste treatment and disposal. Simply put, waste flows head towards places in global systems where there are fewer costs and less red tape.

**E-waste (electrical and electronic waste)** This includes every bit of abandoned electronic and electrical material belonging to computers, CDs, smartphones and printers.

# CONTEMPORARY CASE STUDY: INFORMAL-SECTOR E-WASTE PROCESSING IN GHANA

Ghana receives 200,000 tonnes of imported electronic waste annually, of which 70 per cent are second-hand goods, including donations of old computers sent to schools by charities (but around 15 per cent of second-hand imports are broken beyond repair).

The settlement of Agbogbloshie, on the left bank of the Odaw River in Accra, is home to around 6000 families who migrated here from the north of Ghana to escape tribal conflict and poverty. Over the years it has developed into a dumping ground for old electrical and electronic products. Each month, hundreds of tonnes of e-waste is broken apart by hand to salvage copper and other metallic components.

■ The method of extracting and recovering valuable materials from old computer circuit boards is highly hazardous (see Figure 5.17). The burning process releases toxic substances into the atmosphere, soils and water, with dire health consequences.

■ Known health problems for children as young as ten include acute damage to the lungs from inhalation of fumes of heavy metals such as lead and cadmium.

■ Toxic wastes, heavy metals and battery acids released into the soil and the surface water have destroyed wildlife in the Odaw River, which used to be an important fishing ground for the neighbouring communities.

▲ **Figure 5.17** Working with e-waste in Accra, Ghana

## Climate change refugees

Climate change is causally linked with global system growth. Accelerated cross-border flows of greatly-increased volumes of materials and wastes (not to mention many more people travelling in polluting aeroplanes and land vehicles) have enlarged humanity's planet-wide carbon footprint. The global problem of heightened carbon dioxide and other greenhouse gas (GHG) emissions is only set to worsen, with China's footprint size not expected to peak and decline until around 2040. Many scientists now view a 2°C rise in average world temperature as inevitable, bringing harmful impacts for vulnerable people and places as a result of the enhanced greenhouse effect.

There is injustice in the way that those societies that may be most affected by predicted changes have often done the least to cause them. For example, desertification and the extension of arid conditions in sub-Saharan Africa could adversely affect agriculture and bring food insecurity to regions like Same District in Tanzania and Kitui District in Kenya. UN agencies estimate nearly 10 million people from Africa, South Asia and elsewhere have already migrated or been displaced by environmental degradation, weather-related disasters and desertification in the last 20 years.

 **KEY TERM**

**Food insecurity** When people cannot grow or buy the food they need to meet their basic needs.

- The UN predicts a further 150 million vulnerable people may have to move in the next 50 years and has identified 28 countries now at extreme risk from climate change. Of these, 22 are in Africa. The majority of people displaced by changing weather patterns and sea level rise will be extremely poor.
- More people may become climate change refugees on account of rising sea levels. Dozens of islands in the Indian Sunderbans region are being regularly flooded, threatening thousands. Gravely at risk of rising sea levels, the Maldives is located southwest of India and Sri Lanka. It consists of a chain of 1190 low-lying islands that are surrounded by the waters of the Indian Ocean. Bangladesh is the most vulnerable large country, with 60 per cent of its land less than 5 m above sea level. Villagers in flooded areas will have no option but to migrate; many will go to the slums of capital city Dhaka.

 # Evaluating the issue

▶ *Assessing the efforts of global players to tackle local injustices*

## Identifying possible contexts, criteria and themes for the assessments

This chapter's closing debate assesses ways in which different global players (alternatively called actors or stakeholders) have tried to eradicate or ameliorate some local injustices within global systems. The assessment focuses on the efforts of three categories of global player.

- *Intergovernmental organisations.* IGOs (see page 50) play a very important role. They create new rules, agreements, frameworks and legislation intended to deliver greater socio-economic and environmental justice.
- *TNCs.* As previous chapters have shown, global corporations sometimes bear responsibility for unjust outcomes including poor treatment of workers, environmental degradation, cultural imperialism, tax avoidance and the growth of extreme wealth divides. However, businesses often try to effect positive changes too (page 151 explored

actions taken after the Rana Plaza disaster; page 124 focused on affirmative actions).
- *Non-governmental organisations.* NGOs such as Amnesty International, Oxfam and ActionAid play a critical role in uncovering and raising awareness about economic, social or environmental injustices linked with how global systems work. These non-profit organisations have sometimes pressured state governments and TNCs into working harder to mitigate injustices.

The assessment also requires that we think about different categories of injustice, some of which may be more intractable than others and can only be tackled when different players work together in partnership. As this chapter has shown, there are economic, social and environmental issues to address (see Table 5.5).

- *Economic injustices* arise when financial flows in global system operate in unfair ways. Foreign direct investment into industrialising countries creates new employment, but workers are not always

| Human trafficking (see page 154) | The United Nations Convention against Transnational Organized Crime was signed in 2000. The European Parliament has recognised the gender dimension of human trafficking, stating that: 'Data on the prevalence of this crime show that the majority of its victims are women and girls ... forced into commercial sexual services.' The main EU instrument for fighting trafficking in human beings is Directive 2011/36/EU, adopted in 2011. |
|---|---|
| Over-tourism (see page 157) | The Network of Southern European Cities against Tourism (SET) is a new international movement that lobbies city authorities to use taxes and laws strategically to foster high-spending low-impact tourism rather than mass groups. SET advises tourists to (i) avoid 'top ten' TripAdvisor destinations and visit less well-known places instead and (ii) always behave respectfully (because everywhere is someone else's home place). |

▲ **Table 5.5** How global players have tried to tackle the injustices of human trafficking and over-tourism (topics covered earlier in this chapter)

paid fairly. And although flows between richer and poorer countries are bi-directional (see Figure 5.18), poverty has nevertheless persisted over time in the DRC (see page 160) and other low-income developing countries. Extractivism and resource curse theories help us understand why this happens.

- *Social injustices* have sometimes arisen or worsened because of financialisation (see page 146), leaving poor and vulnerable groups without vital services, homes or land.
- *Environmental injustice* results from land grabs. Outside forces may appropriate natural resources in ways which at best disenfranchise and at worst are injurious to indigenous people.

When analysing injustice, we should think critically too about the varying timescales over which it manifests itself.

- The 'Rhodes Must Fall' campaign is a reminder of historical injustices imposed on the global south by European powers during the colonial era (see page 145). In some people's view, adequate compensation has yet to be offered to societies whose ancestors were enslaved or robbed of their land.
- Equally, new forms of injustice constantly arise as a result of 'shrinking-world' advancements in digital technology. 'Fake'

data flows can have real-world unfair outcomes, for example, if an election result is influenced by lies spread on social media. Victims of online libel or the unpleasant 'revenge porn' phenomenon may struggle to get the content permanently removed from the internet; this failure to respect their 'right to be forgotten' is one of the latest injustices to develop within global systems.

## Assessing TNC efforts to tackle supply-chain injustices

Large businesses increasingly accept the need for corporate social responsibility. As Chapter 3 explained, the largest TNCs have thousands of suppliers they outsource work to. This increases the risk of (i) the violation of workers' human rights and (ii) valuable brand products becoming linked with workforce exploitation.

Many of the TNCs used as case studies and examples in this book have strict codes of practice which prohibit worker exploitation in (i) the offshored facilities they actually own and (ii) their first tier of outsourcing suppliers. A code of conduct guarantees certain rights for employees and may cover legal areas such as health and holiday benefits, maximum working hours or trade union membership rights. One particular success is the Accord on Fire and Building Safety in Bangladesh

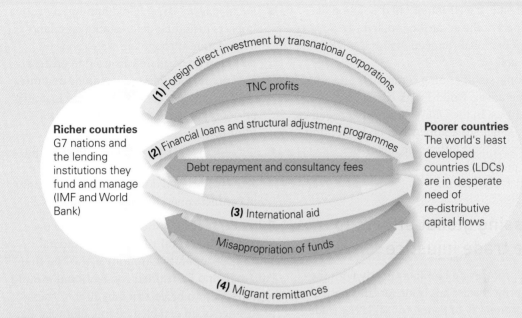

**(1)** Foreign direct investment by transnational corporations

TNC profits

**Richer countries**
G7 nations and
the lending
institutions they
fund and manage
(IMF and World
Bank)

**(2)** Financial loans and structural adjustment programmes

Debt repayment and consultancy fees

**Poorer countries**
The world's least
developed
countries (LDCs)
are in desperate
need of
re-distributive
capital flows

**(3)** International aid

Misappropriation of funds

**(4)** Migrant remittances

▲ **Figure 5.18** Bi-directional financial flows transfer money both ways between 'core' and peripheral' regions within global systems (see page 87). Views differ on whether the net outcome of these flows for poorer countries is growth and development or persisting poverty and injustice

which followed the 2013 Rana Plaza disaster. Over 220 TNCs have now signed up (see page 151). By 2018, inspections had led to safety improvements for more than 1600 higher-tier factories.

However, injustices still occur in many global supply chains. Chapter 3 described how Apple learned that workers for its third-tier supplier, Lianjian Technology, were poisoned by a chemical cleaning agent in 2011. More recently in 2018, allegations surfaced that Quanta Computer, a Taiwanese supplier, illegally employed students in the Chinese city of Chongqing to make Apple Watches (the students were told they would not be allowed to graduate unless they participated). Reports such as these raise question about how far TNCs will ever be able to eradicate abuses and injustices from lower tiers of their complicated supply chains.

## Forced compliance for TNCs

The US government's Dodd-Frank Wall Street Reform and Consumer Protection Act of 2010 is

an example of forced compliance for businesses. This 'top-down' law made it illegal for US-registered TNCs to use **conflict minerals** thought to have originated in the DRC. Dodd-Frank requires US companies to find out where their 3T (tin, tungsten and tantalum) and gold minerals are from and then disclose whether or not those minerals funded armed groups. Most end-user companies did not know their sources of minerals before Dodd-Frank, so the law has forced them to look deeper into supply chains. As a result, militia groups now find it harder to sell gold and diamonds to fund wars which ruin lives. However, there is an opposing view that Dodd-Frank may have inadvertently worsened poverty and instability in parts of the DRC.

● This is because some TNCs have responded by avoiding exports from the DRC altogether, including legitimate supplies, for fear of becoming associated with conflict minerals. They want 100 per cent risk-free operations.

- As a result, a well-intentioned 'solution' is now part of the enduring problems of poverty and poor global connectivity for the DRC and its people.
- For some small mining companies and co-operatives, Dodd-Frank has been disastrous. In the DRC's South Kivu province, many small-scale miners can no longer find buyers for the conflict-free metal ores they have produced, despite living in what is now a conflict-free region.

## Assessing global efforts to tackle trade injustice

You will probably be familiar already with the general principles of fair trade (or 'alternative trade') as one way of trying to tackle the trade injustices caused by extractivism. The work of the Fairtrade Foundation in particular has been very important for some communities in developing countries. The aim is to give producers a decent fixed price for their goods. If the global price for a particular crop like coffee collapses, Fairtrade farmers will continue to receive a steady income that safeguards their wellbeing. The venture only succeeds if a sufficient number of shoppers in high-income countries are motivated by (i) a belief in trade justice and (ii) genuine curiosity about the provenance (origin) of the goods they buy.

Millions of people working in artisanal and small-scale mining (ASM) routinely risk disease, serious injury and death. ASM miners are often taken advantage of by unscrupulous 'middle men', according to the Fairtrade Foundation and the Alliance for Responsible Mining (ARM). There are three interrelated concerns.

- *Pay, health and safety.* There are over six times the number of accidents in ASM compared with large-scale mining, mainly due to its sizeable labour force – which may include children – and poorer working conditions.

- *Environmental issues.* Gold mining's negative impacts including deforestation and land degradation through air, water and soil pollution (toxic chemicals used to process gold ore include mercury, cyanide and nitric acid). Eighty per cent of all human mercury poisoning is caused by artisanal gold mining.
- *Conflict.* Poverty pushes many people into working in artisanal mines, but some are actively forced to do so by militias operating in conflict zones, notably so in the DRC.

Fairtrade and Fairmined gold has now been available for several years. The Cotapata Mining Co-operative in Bolivia was the first Fairtrade and Fairmined conventional mining organisation to be certified in 2011. Others have since been established in Colombia, Peru and Mongolia.

- To gain certification, artisanal miners in a region must first band together to form an organisation that Fairtrade can deal with directly.
- Each organisation pledges to participate in the social development of their communities by eliminating child labour (under 15) from their locality, introducing protective gear for all miners and recognising the right of all workers to establish and join trade unions.
- They must also use safe and responsible practices for the management of toxic chemicals in gold recovery, such as mercury and cyanide.

There are, of course, drawbacks to alternative trading networks and the scale they can operate at. It is simply not possible for all the world's miners, farmers and other producers to join a scheme offering a high fixed price for potentially unlimited volumes of produce (the entire capitalist world system would fail if this happened). Also, the higher price of Fairtrade products means many shoppers avoid them, especially in times of economic hardship.

## Assessing global efforts to safeguard local biological assets

The World Economic Forum, a global not-for-profit organisation, has backed the Earth Bank of Codes (EBC) project which has begun mapping the DNA sequences of all species in the Amazon river basin. The aim is to make sure indigenous communities gain a future share of any profits deriving from the biological assets of their home place (see Figure 5.19).

- Many valuable medicines and products are already derived from Amazonian species, but large pharmaceutical companies will usually have profited, not indigenous tribes. Yet another unfairness is linked with extractivism.
- To prevent further injustice, the biological assets of Amazonia will now be mapped and

analysed by the EBC project. Should any of the biological data prove valuable in the future, indigenous communities will be guaranteed an income.
- This is also in accord with the UN's Nagoya Protocol, an international treaty, the objective of which is: 'The fair and equitable sharing of benefits arising out of the utilisation of genetic resources, thereby contributing to the conservation and sustainable use of biodiversity.'

## Assessing the importance of players working in partnership

Fighting injustice effectively sometimes requires collaboration between different stakeholders. Figure 5.20 shows global players working in partnership with local groups or individuals to create change. Thanks to a campaign which took this approach, conditions recently improved for

▲ **Figure 5.19** Many naturally-occurring substances found in Amazonia's 'gene pool' act as medicines and remedies. For instance, quinine is a tropical plant extract that has been widely used as a painkiller. If valuable new discoveries are made there in the future, who ought to benefit financially? Local tribes, global pharmaceutical companies or both?

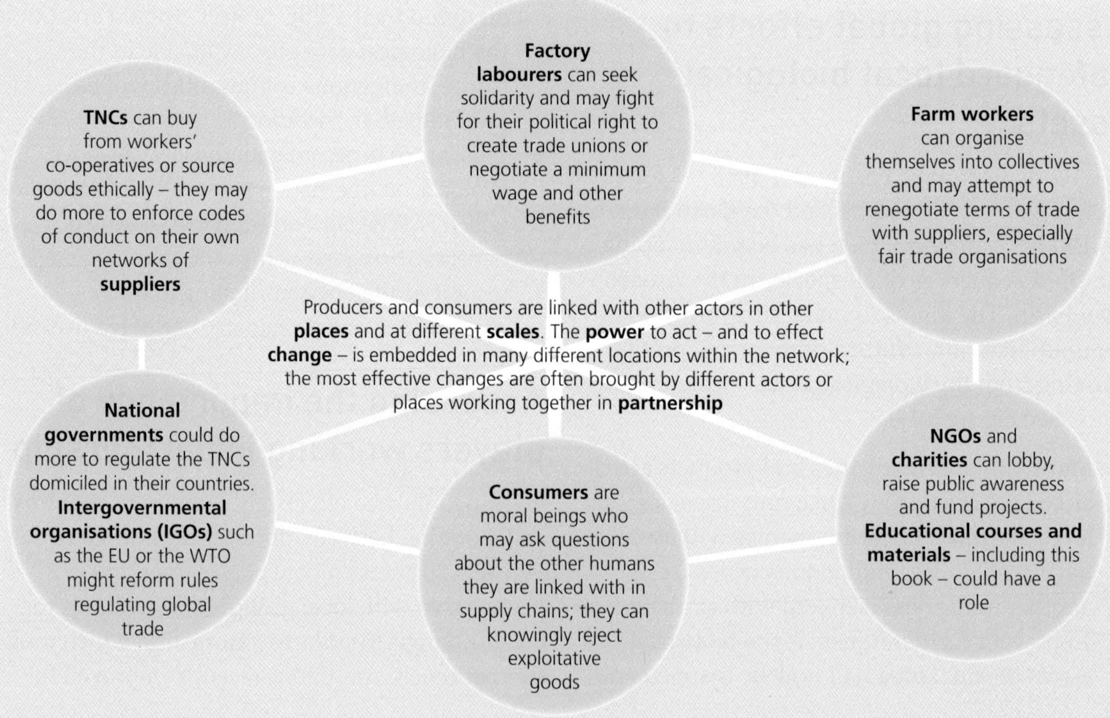

**▲ Figure 5.20** Different stakeholders can work together in an actor-network to bring about positive change and trade justice for local communities

South African women who work in Tesco's global supply chain.

- In 2005, Gertruida Baartman was working as a fruit picker at a South African farm near Cape Town which supplies European supermarkets. At the time, she was paid South Africa's minimum wage of just £97.90 per month.
- The single mother told a newspaper: 'My four children do go hungry but I try my best. I have to pay school fees and sometimes that's a struggle because the fees are high. The school uniforms are expensive for me too and I don't have money to buy them shoes.'
- Two out of three insecure seasonal workers in South Africa are black women like Gertruida. They often lack the same benefits as men, who are more likely to be on permanent contracts.

UK-based NGO ActionAid heard of the plight of South African fruit pickers thanks to its connections with Sikhula Sonke, a women-led trade union of farm workers in South Africa. ActionAid flew Gertruida to the 2007 annual shareholder meeting of the UK-based TNC Tesco in London (see Figure 5.21). She received a standing ovation from the shareholders, who were horrified to hear about conditions at the bottom of their supply chain. After Tesco representatives visited Gertruida's farm, there have been improvements, such as a toilet in the orchard where she works and a reduced pay gap between men and women. In 2012, Sikhula Sonke went on to win a 50 per cent increase in the minimum wage for casual farm labourers in South Africa as part of prolonged strike action.

This study shows how the power to effect change was spread across a network of different individuals and organisations.

- Gertruida lacked financial power and was, potentially, another voiceless labourer. Yet she

▲ **Figure 5.21** Gertruida Baartman attending the Tesco shareholder meeting in 2007

clearly had the drive and determination needed to gain a better outcome for her family once an opportunity was offered.

- ActionAid has limited financial resources which need to be used sparingly and wisely in order to help bring about change. In this example, it set up a meeting between the two polar extremities of the Tesco value chain: labourers and shareholders. Consequently, Gertruida and the shareholders met and gained an understanding of one another.
- Ultimately, the shareholders possess the financial and regulatory power needed to make change (in relation to their own supply chain). Yet they at first lacked knowledge of working conditions among their own subcontractors. Once they learned of the injustice (and risk to their own company's reputation), they took action.

The final outcome is a positive one, though clearly there is a long way to go: pay for South African farm workers is still terribly low by UK standards.

## Reaching an evidenced conclusion

How far have TNCs, NGOs and IGOs effectively tackled different kinds of injustice for particular people and places? Inevitably, the answer is context-specific. Smaller-scale injustices may be easier to rectify, especially when different players collaborate (as happened in the case of Tesco's South African supply chain). Some injustices are more likely to be addressed than others because they are easily visible, while others stay in the shadows. TNC representatives will monitor working conditions in the higher tiers of their supply chains, but poor conditions for other workers may persist, hidden from view, in the lower tiers.

Are some injustices too vast to tackle? Arguably, there remains insufficient acknowledgement of the past injustices suffered by many societies due to colonialism and extractivism. Recognition that indigenous people may have rights over biological assets (as the Earth Bank of Codes project does) is a step in the right direction, but far more could be done. Moreover, some attempts to help have had unintended negative consequences, such as the avoidance of DRC minerals by companies wary of new risks created under the Dodds-Frank Act.

Perhaps the most important way global players can effect change is to use their influence to persuade national governments to accelerate political development processes that will benefit all of a country's citizens. Conditions of work have usually improved in industrialising countries over time (this is a feature of the 'flying geese' phenomenon explored on page 148, but pressure from IGOs, NGOs and TNCs may help speed up changes that address the worst workplace injustices. To some extent, this has been happening in Bangladesh since the Rana Plaza disaster, where TNCs now say they want to work in partnership with the government to improve working conditions for everyone (see also page 151). In 2018, H&M – a major investor in the country – said in a press release that: 'We are committed to using our leverage as one of the biggest buyers in Bangladesh to further improve working conditions, including wages.' Hopefully, this kind of corporate social responsibility will only strengthen over time.

**KEY TERMS**

**Corporate social responsibility** Recognising that companies should behave in moral and ethical ways as part of their business model.

**Conflict minerals** Products of mining industries sourced from conflict zones whose production may have involved slave labour.

# Chapter summary

✔ The concept of injustice encompasses a wide spectrum of unfair outcomes for people and places. There are economic, social and environmental categories of injustice (though in reality these often overlap).

✔ Global systems have given rise to numerous economic injustices, often linked with the financialisation of development and the emergence of a new international division of labour. The spread of neoliberal values has resulted in a tendency to measure everything (from infrastructure to ecosystems) in terms of economic worth or profitability, but this does not always lead to the just and fair treatment of individuals and societies.

✔ Injustice is both a cause and an effect of international migration. Refugees who suffered greatly before fleeing their homes often find that life remains hard even when they have found relative safety. Some migrants are victims of modern slavery and human trafficking. International migration and tourism in global hubs (world cities) is associated with a range of housing and employment market injustices for local communities.

✔ Land grabs and extractivism are forms of environmental injustice which have persisted over time in many countries. Indigenous people have often gained no benefit from the exploitation of raw materials and biological assets found in their home places. Instead, external players have profited from the extraction of these resources and indigenous people have sometimes suffered as part of the process.

✔ Other forms of environmental injustice include the impact of transnational waste movements and global climate change on vulnerable societies in low-income countries.

✔ Attempts to tackle different kinds of injustice have been made by global players (TNCs, IGOs and NGOs), sometimes working in partnership with national governments and local communities. Although some actions have been successful, developmental and technological changes continue to create new injustices for different countries or societies.

# Refresher questions

1 Outline examples of injustice created by global systems for: farm workers; factory workers; call centre workers.

2 Using examples, explain why injustices for factory workers may lessen as a country continues to develop.

3 Explain the causes and consequences of the Rana Plaza disaster in Bangladesh.

4 Using examples, suggest why modern slavery is a difficult problem to solve.

5 Explain how international migration can lead to housing challenges for local communities in global hub cities. Why does tourism make the problem even worse in some cities?

6 What is meant by the following geographical terms? Financialisation; extractivism; resource curse.

7 Using examples, explain why the home places of indigenous people are often vulnerable to land grabbing by external players.

8 Suggest possible costs and benefits for poorer societies created by (i) transnational waste movements and (ii) a warming climate.

9 Describe the strengths and weaknesses of one attempt to tackle an injustice created by global systems.

# Discussion activities

1 As a whole-class activity, compare the concepts of inequality and injustice. Think of possible examples that highlight the differences between the two ideas. Ideally, try to identify inequalities and injustices using examples at different spatial scales. For example, you could start with a localised context, such as an office where not all workers get the same pay. You could then move up to regional and national contexts, for example university fees are unequal in Scotland and England. Is this fair?

2 In groups, discuss the difficulties that could arise when trying to define what is meant by an area's 'indigenous population'. How many generations must a society live somewhere before they are viewed as an indigenous people? Other possible questions to discuss include the following:
   - What happens if there are two or more competing claims from different ethnic groups, and how could any dispute be settled fairly?
   - What rights should local people have over their home place or landscape? In the USA, citizens can profit from oil and gas found below their land, but underground mineral resources are owned by the Crown Estate in the UK. Is it fair that different rules apply in different places?
   - Do you agree with the objectives of the Earth Bank of Codes (EBC) project (see page 167)?
   - Some people call this 'the century of migration'. Should we be worrying more about the rights of migrants, rather than the rights of indigenous people?

3 In pairs, review the case study of Tesco's South African supply chain (see pages 168–69). Who was the most important player in this story, and why?

4 In groups, discuss the relative power and influence that TNCs, NGOs, IGOs and national governments have over the way global systems work. Who has the greatest power to tackle different types of injustice, and why?

# FIELDWORK FOCUS

This chapter's focus – injustice in global systems – could serve as a good independent investigation focus. Possible issues to research include over-tourism, recycling waste flows and refugee movements

**A** *Investigating the local impacts of global flows of tourism.* If you live in close proximity to a major city such as London or Liverpool, then this is a relatively straightforward topic to adopt. There are plenty of opportunities to collect primary quantitative data (pedestrian counts, interview data) and qualitative data (photographs and interview transcripts). You can collect secondary data about house prices and visitor numbers which may help create a fuller 'picture' of the costs of over-tourism for locals. There are models and theories which can be used to underpin the work, including carrying capacity measures.

**B** *Investigating how recycling waste flows connect local households with other places in global systems.* Local authority recycling schemes have led to increased volumes of household waste being sent overseas for recycling. You could devise a way of quantifying the volumes of recycling waste generated by a particular local neighbourhood (e.g. you could visit a sample of homes and ask people how many bags they put out each week; or carry out an actual survey on 'bin morning'). You would need to approach your local authority to find out more about where the waste is sent to (this could be a face-to-face, telephone, or email interview). If you want to maintain a focus on 'injustice' then you would need to carry out secondary research about the impacts of recycling on the overseas communities who process the waste.

**C** *Investigating the experiences of refugees living in the UK.* In theory, this is a very interesting issue to research; but in practice, it could be difficult to carry out. As part of your planning you would need to think very carefully about:

- how to find people to interview (not necessarily an easy thing to do!)

- what questions to ask (and the kind of data the interviews would generate)

- whether you feel it would be entirely ethical to ask people about the hardships they may have endured as refugees.

# Further reading

Burgis, T. (2016) Ethiopia: the billionaire's farm. *Financial Times*, 1 March. Available at: https://ig.ft.com/sites/land-rush-investment/ethiopia.

Goldacre, B. (2013) *Bad Pharma: How Drug Companies Mislead Doctors and Harm Patients.* New York: HarperCollins.

Mawdsley, E. (2016) Development geography 11: Financialization. *Progress in Human Geography*, 1–11.

Rice, X. (2007) The water margin. *Guardian*, 16 August. Available at: www.theguardian.com/business/2007/aug/16/imf.internationalaidanddevelopment.

Tourtellot, J. (2017) Overtourism plagues great destinations. Available at: https://blog.nationalgeographic.org/2017/10/29/overtourism-plagues-great-destinations-heres-why.

# Global systems futures

Previous chapters explained how global flows have brought development, interdependence, inequality and injustices to people and places. In turn, these outcomes have given rise to challenges that may now threaten the future of global systems. This final chapter:

- explores rising tensions between nationalist movements and supporters of 'business as usual' globalisation
- examines the vulnerability of global systems to economic, climatic, technological and demographic disruptions
- evaluates the view that a new era of deglobalisation has begun.

## KEY CONCEPTS

**Deglobalisation** The idea that the world may be experiencing decreased economic integration of countries and reduced cross-border movement of goods, services and capital. Non-economic dimensions of deglobalisation include weakened global governance and increased opposition to the cultural exchanges brought by global migration, media and social networking. Deglobalisation is associated with (i) a global-scale economic slowdown due to problems with the existing system itself and (ii) new political movements that aim to stop or slow different global flows.
**Globalism** The belief that global systems should be encouraged to keep growing. Opponents of globalisation reject globalism as an ideology.
**Nationalism** An umbrella term for a spectrum of new 'populist' political movements which reject the 'globalist' philosophy or significant aspects of it. Typically, new nationalist movements in developed countries demand that the interests of their own country must be more clearly put first, ahead of global issues or rules. State governments may respond with trade or migration barriers.

# 1 Resistance to global systems and flows

▶ *Why is there increasing opposition to globalisation in many higher-income countries?*

Globalisation can be thought of a process, as Chapter 1 explained. Figure 6.1 (overleaf) shows how it has sometimes accelerated or decelerated in response to large economic or political 'shocks'. Aside from these short-term

**Global systems growth phases**

**1940–50s**
Establishment of the World Bank, International Monetary Fund (IMF). The post-war Bretton Woods Conference provides the blueprint for a free-market non-protectionist world economy. Aid, loans and other assistance become available for countries prepared to follow a set of neoliberal guidelines written by powerful nations.

**1960–70s**
Heavy industry in developed countries loses ground to rising production in Asia, including Japan, South Korea and the emerging Asian Tiger economies. Unionised labour costs push up the price of production for Western shipbuilding, electronics and textiles. Elsewhere, soaring petrodollar profits for Middle East OPEC nations signal that the UAE, Saudi Arabia and Qatar are becoming new global hubs.

**1980–90s**
By now, China has begun economic reforms and opens its economy (India soon follows). Financial deregulation in major economies like the UK and USA brings a fresh wave of globalisation, this time involving financial services. Established powers strengthen their regional trading alliances, including the European Union (EU, 1993) and the North American Free Trade Agreement (NAFTA, 1994).

**2000–10s**
Growth remains slow following the global financial crisis. Many countries slip in and out of recession, including Russia and Brazil. Despite slower growth, China overtakes the USA to become the largest economy by purchasing power parity (PPP). Global internet and social networking use reaches new record levels. Experts struggle to tell whether globalisation is increasing, pausing or retreating.

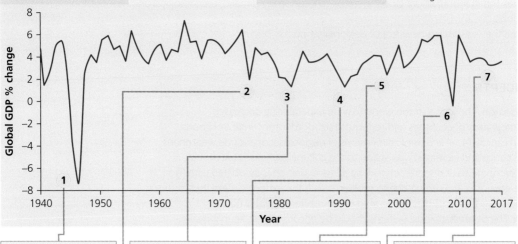

**Major shocks to global systems**

**(1) The Second World War** (1939–1945), when 60 million people died.

**(2) The first OPEC Oil Crisis** of 1973 pushes Western industry over the edge. Rising fuel costs trigger a 'crisis of capitalism' for Europe and America, whose firms begin to step up outsourcing and offshoring of their production to low labour-cost nations.

**(3) The second oil shock** linked with Iraq's invasion of Iran.

**(4) The First Gulf War** (Iraq's 1991 invasion of Kuwait leads to rising oil prices) is followed by the **collapse of the Soviet Union**, which significantly alters the global geopolitical map, leaving the USA as the only true superpower.

**(5) The Asian Financial Crisis** is short-lived but signals the possible risks of financialisation.

**(6) The global financial crisis (GFC) of 2008–09** reveals major flaws in the globalised banking sector. Global GDP falls for the first time since 1945. Problems in Greece and Portugal escalate into the **Eurozone crisis**.

**(7) A continuing period of global instability**. In many countries, popular opposition to migration and free trade is on the rise. Russia annexes Crimea (2014). Daesh (ISIS) and the war in Syria bring new chaos to the Middle East. The UK votes to leave the EU (2016). Donald Trump wins the US presidency (2016), quits the Paris Agreement and raises tariffs.

▲ **Figure 6.1** A timeline showing percentage growth in global GDP, 1940–2018

interruptions, though, the 'roll-out' of globalisation has continued at a relatively steady pace. Until recently, most national governments outwardly agreed that increased global connectivity and interdependence were, on balance, both *inevitable* (for example, because of technological forces) and *desirable* (because of resulting global development and growth). Some writers prophesised an 'end to geography' insofar as nation states would become irrelevant in a 'world of flows' (see page 42). Accelerated interconnectedness and interdependence between societies signalled a new 'shrinking' and 'borderless' world, and the erosion of differences between places.

In a famous 1999 speech, UK Prime Minister Tony Blair made the following forecast for the twenty-first century:

*'We are all internationalists now, whether we like it or not. We cannot refuse to participate in global markets if we want to prosper. We cannot ignore new political ideas in other countries if we want to innovate. We cannot turn our backs on conflicts and the violation of human rights within other countries if we want still to be secure.'*

The position of successive British governments, in common with other high-income countries, has been to embrace globalisation while seeking to tackle some of the economic, social and environmental injustices that global systems create, both domestically and in other countries. In particular, the UK has (i) long maintained a sizeable international aid budget to assist developing countries and (ii) supported numerous UN environmental agreements.

Independently of government, many citizens act in ways they hope will mitigate the worst effects of global systems. Some buy fair trade products (see page 166). Others may participate in fundraising or other campaigns in support of trade justice or environmental causes (see Figure 6.2). Supporters of these 'progressive' causes are not necessarily opposed to globalisation *per se*. They may well believe the best way to help people and places in developing countries is with stronger global rules and increased (not lessened) international co-operation. *In short, they trust that the same global systems that create problems also provide the best means of solving them.*

Progressive causes have been championed by, among others, the UK Labour Party and the US Democratic Party.

▲ **Figure 6.2** Some 'anti-globalisation' demonstrations are linked with progressive political movements. Around 40,000 people demanded improved trade justice for developing counties at the 1999 WTO conference in Seattle. Outside the 2015 Paris climate change conference (shown here), campaigners dressed as polar bears and penguins called for even greater global co-operation on climate change. In these and other examples, protestors see themselves as 'concerned global citizens'. They aim to reform global systems. In contrast, some newly emerging nationalist movements advocate a partial retreat from, or more complete rejection of, global systems

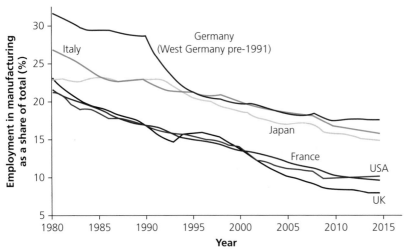

**▲ Figure 6.3** Employment losses in manufacturing have brought hardship to working communities in all of these countries. *FT Graphic; Source: European Commission*

- As a result, these left-wing political parties typically attract voters who are more concerned with social and environmental justice issues.
- However, left-wing politicians have sometimes struggled to find a clear policy position in relation to globalisation and trade. This is because deindustrialisation has brought hardship to poorer and more vulnerable communities in both countries (see Figure 6.3). Some left-wing politicians may therefore be sympathetic to the idea of trade barriers if it helps prevent factory closures in their own constituencies.

## The resurgence of nationalism

While 'progressive' political movements want to reform globalisation, new nationalist movements go one step further by advocating a partial retreat from global systems. As Table 6.1 shows, there is growing evidence of people and political parties voicing their dislike of the interdependence which globalisation fosters. To judge by their actions, they prefer *in*dependence. Social commentators say that twenty-first-century politics is increasingly made up of two competing perspectives: globalism and nationalism (see Figure 6.4).

**'Globalists' are more likely to support or value:**
- intergovernmental agreements and institutions
- international-mindedness and open borders
- cultural diversity and changing identities
- making perceived global injustices a priority for action (e.g. refugees, 'sweatshops' and global environmental issues)
- the concept of global citizenship

**'Nationalists' are more likely to support or value:**
- political sovereignty (national independence)
- closed borders and barriers to trade and/or migration
- cultural homogeneity (cultural uniformity)
- making perceived local injustices a priority (e.g. concerns with employment, housing and immigration in their locality)
- their own national citizenship above that of others

**▲ Figure 6.4** Two parallel systems, globalism and nationalism

| UK | At the time of writing, the UK is expected to leave the EU (while perhaps remaining in a customs union). |
|---|---|
| USA | During 2018, President Trump imposed new import taxes worth US$200 billion on (mainly Chinese) goods. He has also called for a wall on the US–Mexico border and the reshoring of American industry. |
| Poland | Prime Minster Mateusz Morawiecki refused to sign a new UN migration agreement. Austria, Hungary and other countries also turned their backs on the UN Global Compact for Migration. |
| France | Support continues to grow for Marine le Pen's Front National (a nationalist party). |
| Germany | In 2018, nearly 1.5 million refugees were living in Germany – three times more than in any other EU country. But many citizens oppose their government's *Willkommenskultur* (welcoming culture). |
| Italy | The anti-establishment Five Star Movement won the most votes in the 2018 general election. |
| Brazil | The 2018 presidential election was won by Jair Bolsonaro, a self-professed admirer of Trump. |
| Russia | The 2014 annexation of Crimea showed Russia's disregard for a rules-based international order. Sweeping sanctions were introduced against Russia by the EU and USA. |
| Venezuela and Bolivia | Several South American governments have taken back control of their own energy supplies from foreign TNCs such as BP, ExxonMobil and Repsol. This is called 'resource nationalism' (see page 108). |

▲ **Table 6.1** Signs of the resurgence of nationalism (and globalism in retreat) in selected countries. Nationalist parties and politicians often oppose immigration; some reject multiculturalism entirely. However, only a minority have an extreme position; most people are far more moderate and reasonable in their demands for 'taking back control'

Globalism is often characterised as an ideology followed by 'social elites' (professional, university-educated people). In contrast, nationalism is portrayed as being popular with 'ordinary working people'. The reality is more complex because views can also polarise according to people's age, ethnicity and where they live (urban or rural). Also, the two value systems *are not necessarily oppositions* though they are often portrayed that way. It is possible to have a global outlook while remaining deeply patriotic, for instance. There is, unfortunately, a widespread tendency among those on both sides of the debate to portray one another unfairly.

- In their speeches, some political leaders, including Trump and the UK's Nigel Farage (ex-leader of the UK Independence Party, or UKIP), have suggested that so-called globalists have for decades conspired to undermine national governments (see Figure 6.5). This idea – that globalism is an 'elite project' favoured by the rich – has spread widely in mainstream and social media alike.
- Meanwhile, the so-called globalists have not always done themselves favours by 'talking down' to nationalists. While campaigning for the US Presidency in 2016, Hillary Clinton derided Trump supporters as 'deplorables', further characterising them as 'racist, sexist, homophobic, xenophobic, Islamophobic'. This upset many people who argued that she was out of touch with quite reasonable worries about immigration and trade expressed by many millions of ordinary US citizens.

When a majority of UK voters chose to leave the EU in 2016, pro-European politicians including Prime Minister David Cameron (who had called the

> Globalisation creates economic policies where the transnationals lord over us, and the result is misery and unemployment.

Evo Morales, President of Bolivia, 2006

> The rampant globalisation that is endangering our civilisation ... It puts on me a huge responsibility to defend the French nation, its unity, its security, its culture, its prosperity and its independence.

Marine Le Pen, leader of France's Front National party, 2017

> A globalist is a person that wants the globe to do well, frankly, not caring about our country so much. And you know what? We can't have that ... You know what I am? I'm a nationalist.

Donald Trump, US President, 2018

▲ **Figure 6.5** In the early 2000s, leaders of some developing countries, such as Bolivia, often spoke out against globalisation. More recently, nationalism has become part of everyday political life in leading G7 economies too

referendum on membership, expecting the 'remain' side to win comfortably) were stunned by the unexpected result.

- Attracted by the 'leave' campaign's promise of stricter migration controls and the restoration of national sovereignty, British people voted to abandon their economic, political and demographic union with 27 neighbour states.
- The *Financial Times* newspaper described the referendum result as a 'roar of rage' by people who feel 'alienated by globalisation'.

## Explaining nationalism and the retreat from globalisation

Nationalism is not a new phenomenon; it was a powerful force in the early decades of the twentieth century, for instance. The new nationalist political forces that have sprung up or strengthened in many countries since the GFC are alternatively described as 'populist' movements. This means they hinge on a set of 'everyday' concerns which a critical mass of lower- and middle-income people are believed to share. The issues responsible for resurgent nationalism include the following.

- *The global shift of manufacturing work away from developed countries.* Although many jobs have been lost to automation, offshoring has played a role too. In the USA, there is a belief among many voters that 'American jobs' have been lost unfairly to Mexico and China. There is a further feeling that cheap imports from those countries threaten what manufacturing remains in the USA.
- *Membership of intergovernmental organisations (IGOs).* In some people's view, intergovernmental organisations pose a threat to the independence of the nation-state. In the minds of many Brexit voters, EU membership (along with the European Parliament and the European Convention on Human Rights) had robbed the UK of national sovereignty. Many citizens in other EU states feel the same way.
- *International migration.* A weight of evidence suggests economies benefit from the arrival of young and ambitious migrants. However, many people remain more concerned with what they see as a threat to their country's community cohesion. Political parties and organisations opposed to the free movement of people (including the passport-free Schengen Agreement) can be found throughout the EU and support for them is growing. Immigration is 'globalisation made flesh', according to the economist Martin Wolf.

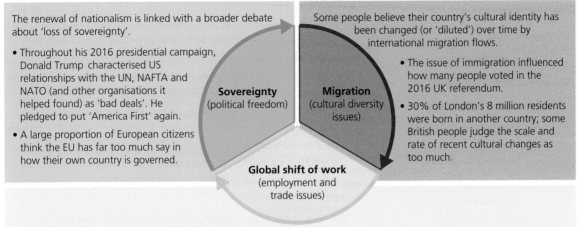

The renewal of nationalism is linked with a broader debate about 'loss of sovereignty'.

- Throughout his 2016 presidential campaign, Donald Trump characterised US relationships with the UN, NAFTA and NATO (and other organisations it helped found) as 'bad deals'. He pledged to put 'America First' again.
- A large proportion of European citizens think the EU has far too much say in how their own country is governed.

**Sovereignty** (political freedom)

**Migration** (cultural diversity issues)

**Global shift of work** (employment and trade issues)

Some people believe their country's cultural identity has been changed (or 'diluted') over time by international migration flows.

- The issue of immigration influenced how many people voted in the 2016 UK referendum.
- 30% of London's 8 million residents were born in another country; some British people judge the scale and rate of recent cultural changes as too much.

A view has developed, rightly or wrongly, that globalisation is a plot which benefits the shareholders of TNCs but deprives working people of jobs lost to outsourcing and offshoring (see Chapters 1 and 3).

- According to this narrative, world trade has benefited 'elites' in the southeast of England and Washington, DC, but not 'ordinary folks' living in northern England or the US Midwest.
- Some people argue that competition from China's businesses is unfair because of the way that the Chinese government provides financial support for its industries, thereby lowering production costs.

▲ **Figure 6.6** New nationalist (alternatively called 'populist') political movements in developed countries typically share three overlapping concerns: sovereignty, migration and the global shift of work

In summary, Figure 6.6 portrays nationalism as a **nexus** comprised of these three overlapping concerns, namely (i) sovereignty, (ii) migration and (iii) global shift.

## The perception of injustice

Chapter 4's plenary analysis of inequality (see pages 133–140) showed how globalisation's 'winners and losers' are not kept neatly apart by national borders. Instead, a more complex pattern exists. Figure 6.7 offers more corroborating evidence of this. It shows the *very* high proportions of people in developed countries whose incomes remained stagnant or fell between 2005 and 2014. But during this same period, incomes rose upwards significantly for the new middle classes of China, Brazil and other emerging economies (albeit from a very low starting point).

Rightly or wrongly, a significant proportion of the developed-world populations shown in Figure 6.7 see themselves as victims of economic injustice.

- They say their governments were slow to recognise and mitigate the negative impacts of global systems on local places, especially job losses (caused by offshoring and deindustrialisation) and affordable housing shortages. Previous chapters have shown there is some truth in this: property prices in popular cities like London have soared (see page 155), partly as a result of laissez-faire attitudes towards economic globalisation favoured by neoliberal governments and financial institutions (see page 43).

**🗝 KEY TERM**

**Nexus** A series of linked and interconnected things that cannot be understood, or managed, in isolation.

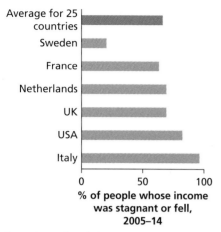

Source: Poorer Than Their Parents? McKinsey Global Institute report, 2018

▲ **Figure 6.7** Around two-thirds of households in high-income countries experienced stagnant or falling incomes between 2005 and 2014, whereas in the decade before this 98 per cent had enjoyed growth

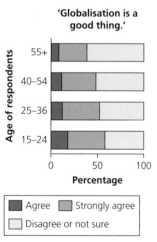

Source: Surveys carried out across the EU in 2017 by Eurobarometer

▲ **Figure 6.8** Belief in the merits or otherwise of globalisation varies markedly with people's age. About 60 per cent of under-25s have a positive view; attitudes differ markedly for the over-55s

- Political and economic freedoms have been engineered for TNCs to invest worldwide and build extensive supply chains, making full use of free-trade zones and tariff-free trade blocs. Unfortunately, the risks that multiply in these systems may go unrecognised by governments until it is far too late: the GFC is the ultimate example of this (see page 91). In the UK especially, a strong argument can be made that social elites profited greatly from the financial systems that eventually triggered the GFC; but the government spending cutbacks that followed had disproportionately negative effects for Britain's poorest and most vulnerable communities.

However, there is an important and sometimes overlooked counterargument to the view that lower-income developed-world populations are 'losers' of unjust global systems. The flow of cheap goods from China to the EU and USA has brought items like the iPhone and flat-screen TVs to 'ordinary' people who might not have been able to afford these same things if they had been made in developed countries with much higher labour costs. By some measures, the cost of living is relatively much cheaper than it used to be thanks to global shift. Equally, some tough working conditions persist at the other end of the chain in so-called 'winner' countries such as India, South Africa (see page 168), and parts of China (see page 150).

Finally, it is important to remember that not *all* of the people with stagnant incomes in Figure 6.7 exhibit antagonism towards global systems. Several important globalist–nationalist 'divides' have been identified by social scientists who have analysed voting patterns in recent European and North American elections.

- Attitudes towards globalisation vary considerably for different age cohorts in EU countries (see Figure 6.8).
- University-educated higher earners were less likely to support Brexit and Trump than people with fewer qualifications and smaller incomes.
- A rural–urban attitudinal divide exists in most countries. Trump was carried to the White House by strong polling in rural states. Support for Brexit was much stronger in rural parts of the UK than in major cities like Cardiff and London. Developing countries are similarly divided: in rural northeast Nigeria, a violent campaign against the Westernisation of Nigerian society is being led by the Boko Haram militia group. In contrast, coastal megacity Lagos is a major hub of the global economy.

## A constant correlation

A 'constant correlation' is how philosopher Michel Foucault described the tension that often arises when an individual becomes part of a 'totality'. In other words, the greater our sense of belonging to something becomes, the more we may come to value what makes us different from everyone else. And as globalisation accelerated in the 1990s and early 2000s, so too did resurgent nationalism become more visible in many places. 'Even as the world enjoys the benefits of unprecedented free movement of goods, people and ideas,' argued historian Simon Schama more recently, 'so it also recoils against those same things.'

# CONTEMPORARY CASE STUDY: RESISTANCE TO GLOBAL SYSTEMS IN INDIA

Complaints from voters in Western countries about globalisation's downsides are relatively new. In contrast, communities in developing and emerging countries often have far longer-standing grievances against the injustices of global systems. Although British colonial rule ended there in 1947, India continues to be influenced by outside global forces in ways some people view negatively.

## Resisting the McDonaldisation of India

Despite McDonald's use of glocalisation to gain acceptance in different markets, including India (see Chapter 2, page 72), the company attracts harsh criticism on the grounds that it is responsible for cultural homogenisation on a global scale. There is concern that local cultures are being 'extinguished' as a result of the global roll-out of uniform (albeit glocalised) services, such as McDonald's restaurants – some people even term it 'McDonaldisation'. Globalisation sceptics are deeply concerned that local cooking traditions will be lost to the onslaught of fast-food menus (often written in English too, thereby accelerating the decline of local languages).

In 2012, McDonald's took the step of opening two entirely meat-free vegetarian restaurants in India. Religious pilgrims to two of India's most sacred spiritual sites found the golden arches of McDonald's waiting there for them.

These vegetarian outlets were developed in Amritsar, home to the Golden Temple, the holiest site of India's minority Sikh faith, and the town of Katra, the base for Hindus visiting the mountain shrine of Vaishno Devi, the second busiest pilgrimage spot in India.

The move was not been popular with some people, however. The Hindu nationalist group Swadeshi Jagran Manch, a branch of the influential Rashtriya Swayamsevak Sangh (RSS), opposed the arrival of McDonald's.

'It's an attempt not only to make money but also to deliberately humiliate Hindus. It is an organisation associated with cow slaughter. If we make an announcement that they're slaughtering cows, people won't eat there. We are definitely going to fight it,' a spokesperson told one newspaper in 2012.

However, McDonald's has since gone from strength to strength in India, serving over 300 million customers in 2017, including those in Amritsar. Burger King, KFC and Dunkin' Donuts all have a growing presence in India too.

# CONTEMPORARY CASE STUDY: THE USA'S CHANGING ROLE IN GLOBAL SYSTEMS

A previous book in this series, *Globalisation* (2011), argued that:

*The nations that have benefited most from globalisation until now have been those with sufficient geopolitical weight to control the terms of their own global interactions with other countries and with TNCs, in ways that yield significant economic and political rewards.*

More than any other country, this claim applies to the USA. It is a true world superpower whose industries (including Apple – page 89; Facebook – page 29) have benefited enormously from globalisation. However, the USA's relationship with global systems is changing because a growing number of its citizens now believe that they *personally* have not benefited from globalisation (see Figure 6.9). President Trump was elected in 2016 sharing those beliefs. Once in office, he began using his power and influence to undermine those very IGOs which the USA was originally a driving force in building, including the UN (and its many agencies), NATO and NAFTA (see page 85).

President Trump arrived on the political stage as a self-proclaimed nationalist (see Figure 6.5) who appeared to see little virtue in concepts such as interdependence and global citizenship. After his election, the USA remained a powerful player in global systems but in changed ways. Under previous post-war leaders, the US government sought to drive global changes (usually to its own advantage) by assuming the role of 'captain of the team'. In contrast, President Trump took a new approach by acting like 'the strongest boxer in the ring'.

To what extent then is the USA still a driving force in global systems?

- In recent years, the USA has successfully developed its own shale gas resources. As a result of increased domestic energy security, there is a lessened need for US overseas geopolitical involvement in the oil-rich Middle East.

- Withdrawal from the Paris Agreement on climate change was symptomatic of increased US disengagement with global systems following the election of President Trump (a self-professed climate change sceptic, he issued an executive order to withdraw in 2017).

- One result of the US retreat from the Paris Agreement is that China has begun to assume a leadership role in climate talks, filling a vacuum left by the USA. In this and other ways, we are seeing a shift from a unipolar world (where the USA was the only superpower driving global systems) to a multi-polar world where power and influence is more evenly spread between Russia, China and the EU, among others.

- Figure 6.10 shows the significant fall in refugees admitted to the USA in 2018. Illegal migrants have sometimes been subject to treatment many people have deemed harsh (in one highly publicised case, children were separated from their parents and put in detention camps near the Mexican border). This marks another retreat from global systems for the USA. Between 1880 and 1930, nearly 30 million new arrivals were registered, including Donald Trump's own grandfather, thanks to the 'open door' attitudes and policies of that era. Over 200 million US citizens alive today are descendants of migrants; but in recent years migration restrictions have been introduced and the coveted US Green Card has become harder to gain.

Will future US governments reverse course, for instance by lowering tariffs again? US citizens are divided over this and other issues, often due to their own differing circumstances.

- In the 2018 midterm elections, some soybean farmers were reported to be unhappy with the damage done to their own export sales by President Trump's new trade barriers with China.

- In contrast, other US citizens hope that future governments will continue the work begun by President Trump: they do not want the USA to resume its role as the driving force behind globalisation. They agree with putting up trade and migration barriers.

- The situation is further complicated by the way some US state governments (notably California) and TNCs (Apple, Facebook and others) maintain a strong global outlook and commitment to international environmental and development goals.

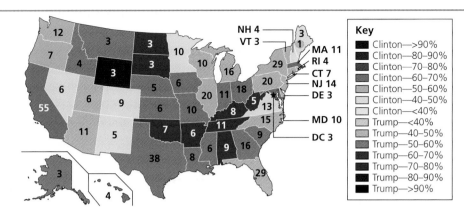

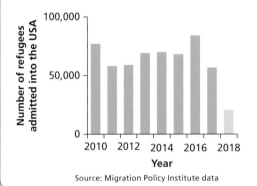

▲ **Figure 6.9** The electoral college votes per state in the 2016 US presidential election. Some people viewed the result as a 'vote against globalisation'. Donald Trump's brand of nationalism proved very popular in rural and deindustrialised states where many voters felt 'left behind'. In contrast, his opponent, Hillary Clinton, polled well in relatively prosperous regions where more people have a 'global outlook' (such as the West Coast, where many global technology companies, including the FANGs, are based)

◀ **Figure 6.10** The number of refugees allowed into the USA was radically reduced in 2018 by the Trump administration

Source: Migration Policy Institute data

 ## Global challenges

▶ *Why may global systems face an increasing risk of disruption in the near future?*

Alongside nationalism, significant forces threaten to disrupt global systems further. Numerous question marks hang over future global economic health.

- Can the global economy ever regain the higher growth rate it enjoyed in the early 2000s, before the global financial crisis (GFC), or has development slowed permanently?
- Will world population keep growing, as demographic models suggest? Meeting the needs of an additional 2 billion people by 2050 could be difficult. Might some existing territorial disputes over land and water escalate into more serious conflicts? Rising affluence in Asia and the Middle East means it may become harder to meet all nations' changing demands.

- Can collective climate change mitigation efforts by the world community safely secure a maximum global temperature rise of 1.5 °C? This value is now identified by the majority of leading scientists as the critical safety threshold.
- Will new technologies help provide solutions for these and other challenges? Equally, what new risks for people and places are created by emerging technologies such as robotics and AI?

This section now works in turn through these propositions.

## The global economic challenge

The 2008 GFC (see page 91) marked the start of a new chapter for global systems. In 2018 – a full ten years later – the GFC's lasting effects were still clearly visible.

- Income growth in the UK remained depressed after the crisis (see Figure 6.11). In 2018, per capita GDP was around 16 per cent smaller than it would have been had it followed the pre-crisis trend. This goes a long way towards explaining many British people's continuing dissatisfaction with the political *status quo*.
- The year 2016 was the fifth consecutive year when global trade did not grow (as a percentage of GDP). International flows of trade, services and finance grew steadily between 1990 and 2007 before collapsing and stagnating. Annual cross-border capital flows have not returned to their 2007 peak of US$8.5 trillion. Container shipping movements have also declined. The Baltic Dry Index – a measure of the price for shipping dry goods such as iron ore and coal – reached a new record low in 2016 (though it later recovered slightly).
- In 2017 and 2018, minor increases in global economic growth were recorded but opinion among economists remains divided over whether or not a lasting slowdown has occurred in global trade and growth compared with previous decades (see Figure 6.11).

One especially important influence on world trade is the way the Chinese economy has matured. Despite being the world's leading economy by some measures, its growth rate more than halved during 2007–18, from 14 per cent to just under 6 per cent. China was globalisation's growth locomotive: slowdown of the world's largest economy has serious implications for everyone. Moreover, this is a permanent rather than cyclical change because China has entered a new and slower phase of economic development. Instead of exporting cheap mass-manufactured goods, China's leaders have

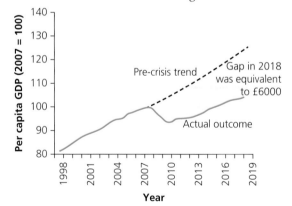

Source: Institute of Fiscal Studies data

▲ **Figure 6.11** UK income growth (indexed values), 1997–2019. Actual changes differ markedly from the projected pathway prior to the GFC. In 2018, per capita GDP was about £6000 per person lower than it might have been had pre-GFC growth trends continued. This suggests the UK economy has shown poor resilience overall

re-focused the Chinese economy on the production of more sophisticated and higher-value consumer items for the country's own domestic markets.

This has, in turn, reduced China's overall demand for natural resources, thereby ending a decade-long global commodities boom or 'super cycle'. Prices fall when markets weaken and in 2016 the prices of iron ore, aluminium, copper, gold, platinum and oil reached their lowest levels since the GFC.

- In large part due to reduced Chinese demand, economic growth in some resource-producing sub-Saharan nations was half what it had been in 2018 compared with a decade earlier, leading several countries (including Mozambique and Zambia) to ask the IMF for further assistance.
- Dramatic oil price falls after 2014 put severe stress on some of the world's oil-producing nations, notably Venezuela and Nigeria, whose economy entered recession again briefly in 2018 (a worrying condition for a country with high youth unemployment whose population will double in size again by 2050.

At a time when many national economies remain relatively weak, it is unfortunate that new nationalist and protectionist political forces have begun chipping away at the global collaborative spirit that helped contain the last financial crisis in 2008–09. The economic rebound of 2010–11 shown in Figure 6.12 was only achieved with a high level of international co-operation, particularly among G7 and OECD members (see page 56). Since then, increased tension in international politics means global systems will most likely prove less resilient should another financial crisis develop.

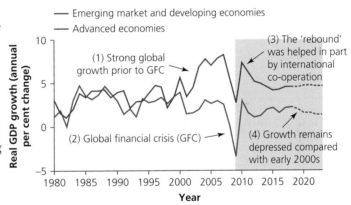

▲ **Figure 6.12** Real GDP growth for emerging and advanced economies, 1980–2020 (IMF historical and projected data). One view is that a 'boom' period lasting from 1980–2007 has now given way to a new phase of permanently reduced growth

## The demographic challenge

Global population changes introduce further uncertainty over the future of global systems and flows. Two important projections stand out in particular.

1 Globally, fertility has fallen significantly to a current average of 2.3 children per woman in 2018. Population growth in Asia and South America has all but levelled off. This is not the case in much of Africa, however, where fertility rates sometimes still exceed six children per woman (in Niger and Mali, for example). Currently, there is no way of knowing how long fertility will remain high in many African countries and regions. That will depend on how strong local resistance to cultural

change remains. For this reason, population forecasts for Africa in 2100 range from 2.5 billion to 4 billion people (the latter represents a quadrupling of the current population). We can predict all African states will experience declining fertility eventually; however, we cannot say *when* this will happen with much confidence. In every scenario, there are significant implications for global patterns of investment, trade and migration.

2 The global **ageing population** issue is important too. In Japan, Iceland, Italy, Australia, Germany and many other countries, life expectancy is 80 or higher. More than 20 per cent of the population in these countries are currently aged 65 or over. In the future, the proportion of elderly citizens in developed countries will grow higher still, while most middle-income countries will also begin to experience the effects of widespread ageing. As we have seen, resistance to immigration is one of the driving forces behind new nationalist movements. Yet it is certainly the case that most developed countries will need to attract more youthful immigrants in the years ahead in order to maintain a relatively youthful population that can maintain economic productivity. Japan's government has long upheld barriers against permanent international migration (see page 48). In 2018, however, Prime Minister Shinzō Abe announced that rules will be relaxed in recognition that Japan is the world's fastest-ageing country.

# ANALYSIS AND INTERPRETATION

Study Figure 6.13, which shows predicted changes in the size of large cities.

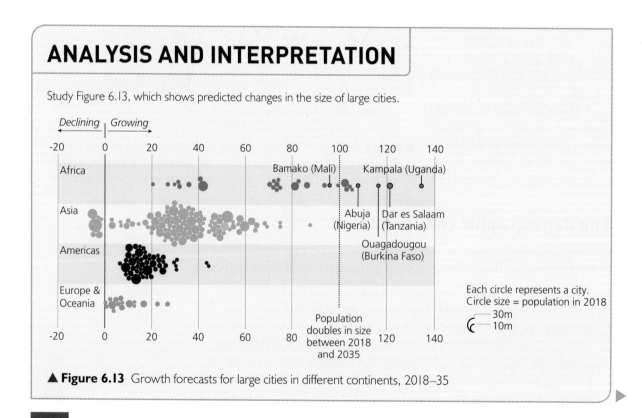

▲ **Figure 6.13** Growth forecasts for large cities in different continents, 2018–35

(a) Estimate (i) the predicted size of Africa's largest city in 2035, and (ii) the number of cities in Africa predicted to more than double in size by 2035.

### GUIDANCE

Make sure you estimate the size of the largest city carefully using the proportional circle scale.

(b) Explain why cities in Europe and Oceania are predicted to grow by no more than 30 per cent whereas many African cities will double in size.

### GUIDANCE

This answer requires you to apply A-level knowledge about global flows while also showing understanding of key demographic principles. Fertility rates are very low now in developed countries. The city growth shown in Europe and Oceania (Australia and New Zealand) can only be attributed to internal and international migration. Most of the cities shown are doubtless located in the EU: a good answer will explore the reasons why cities like Paris and Berlin are likely to keep growing in size because of freedom of movement (a core EU principle). In contrast, growth in most African cities will result from a combination of migration and natural increase (high fertility).

(c) Suggest how the changes shown for Asian cities could affect the volume of **two** global flows connecting Asia with other parts of the world.

### GUIDANCE

Possibilities include flows of raw materials, manufactured goods, services, investment, people (migrants or tourists) or data. For example, you could focus on the increased volume of trade and investment flows that could result from population growth. In order to provide an outstanding answer, try to offer suggestions that are not unduly simplistic and remember that changes could also be contingent on local geographical factors. Do not assume all places will change in the same way. For example, a growing city may attract flows of investment if it is well managed by city authorities (improved transport infrastructure could make a city a good outsourcing base for TNCs, especially if it has a coastal location). However, a poorly managed and increasingly congested landlocked city might not attract new investors.

## The climate challenge

At some point in the future, global fossil-fuel production and use will finally peak before declining. Until then, anthropogenic carbon emissions are expected to rise, as a regrettable by-product of the growth and development which global systems have helped create (see Figure 6.14a). Climate change scientists say fossil fuels must be abandoned – more or less immediately – to prevent dangerous climate change. Figure 6.14b shows what scientists say needs to happen – but this conflicts with the interests of powerful global players, including energy companies and oil-producing nations.

Figure 6.14c shows how global systems have induced a so-called wicked problem composed of many interrelated political, economic, physical and technological elements.

 **KEY TERM**

**Wicked problem** Some challenges and issues are far harder to tackle than others. They cannot be solved easily using conventional decision-making logic or cost–benefit analysis. Instead, they persist as 'wicked problems'.

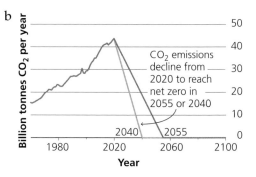

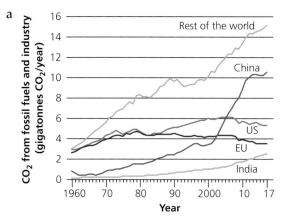

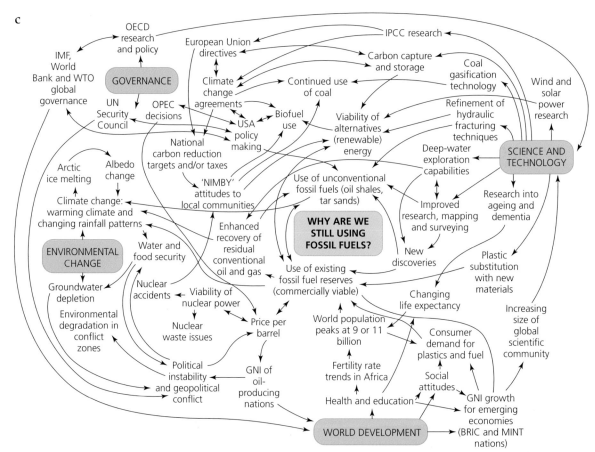

▲ **Figure 6.14** The fossil-fuel dependency of global economic systems presents policy-makers with a wicked problem to solve. These illustrations show that (a) greenhouse gas emissions are projected to rise, (b) immediate emissions cuts are needed order to limit global warming to 1.5°C, but (c) fossil fuel use is currently affected by too many variables. *(a) and (b) are FT Graphics; Source: Global Carbon Project*

- Climate change has arisen from the overlap of two complex systems: the atmospheric system and global economic system. Both are composed of many interdependent parts whose operations are not fully understood.

- A range of physical environments and societies are now threatened in multiple and interconnected ways.
- Some people and places will be affected far more than others, resulting in an uneven political impetus to act.
- The issue of fossil-fuel use cannot be tackled adequately without the co-operation of many players at varying geographic scales, including national governments.
- Any proposed solutions may have complex effects that, in time, create further problems.

In summary, the complexity of global systems makes effective climate change mitigation an elusive goal, as shown by the complex nexus of interconnected issues and actors represented in Figure 6.14c.

# Technological changes, challenges and fixes

As we have seen, technology has been a driving force behind globalisation. Transport and communications growth improvements have often brought new growth and development: new call centre jobs in India and the Philippines support this argument, for example. But each new wave of technology destroys jobs too – think of how the automation of first farming and later manufacturing decimated employment in developed countries during the twentieth century.

## Emerging technological changes and challenges for global systems

Another big 'question mark' over the future of global systems is the extent to which new technologies will continue to create jobs in developing countries. Chapter 4 looked at how significant average income rises among factory workers have sometimes brought development to middle-income countries including China, India, Indonesia, Brazil, Mexico, Nigeria and South Africa. Industrialisation has long been seen as a first step towards prosperity for agricultural nations, based on the previous experiences of developed countries.

The problem now, according to Harvard economist Dani Rodrik and a growing number of financial commentators who use his data, is that peak manufacturing employment may have already arrived in some countries mere decades after the first arrival of factories. Moreover, this is happening at a time when many populations are still growing, albeit at slowing rates. Rodrik's data are shown in Figure 6.15.

- Emerging economies are not gaining the same high percentage of manufacturing employment that some developed countries once did. What's more, the onset of 'peak industrialisation' (when manufacturing employment reaches a maximum, when measured as a share of all work) begins far sooner in Rodrik's sample of emerging economies than it did in most developed countries.

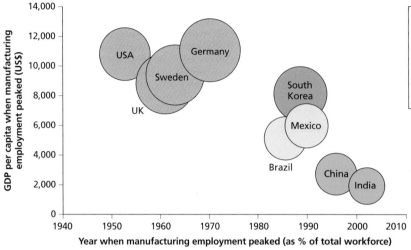

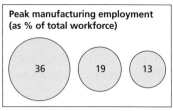

◀ **Figure 6.15** Early industrialising nations like Germany achieved a peak manufacturing employment share in excess of 30 per cent, but emerging economies have done far less well, with Brazil peaking at 16 per cent and India at just 13 per cent

🔑 **KEY TERM**

**Premature deindustrialisation**
An early fall in the relative importance of manufacturing employment, potentially denying a country the full social benefits that industrial development eventually brings.

- The shorter gap between the onset and peak of industrialisation in emerging economies means that the income per capita gains of this period of economic development have been lower than they were for developed countries.

What accounts for this **premature deindustrialisation**? One argument is that advancements in AI and robotics are 'chipping away' vigorously at the numbers of workers needed in factories, even in countries where wages are low.

- The rise of so-called 'SEWBOTS' is particularly worrying for Bangladesh, where large numbers of people work making textiles. Developed by SoftWear Automation, the SEWBOT is a fully automated and highly intelligent sewing robot.
- Taiwanese electronics group Foxconn, which makes Apple iPhones, recently announced it aims to replace one-third of its workforce with robots. Much of its work is carried out in China where average hourly wages had risen to US$3.60 in 2018, meaning there are clear cost savings from adopting AI technology.

The premature deindustrialisation of emerging economies brings challenges and opportunities for global systems.

- On the downside, the assumption that manufacturing will provide future work for Africa's growing population (see page 186) may prove false and a demographic dividend may become wasted in many countries. Pessimists in turn fear increased instability and misrule in countries with high levels of youth unemployment.
- On the other hand, say optimists, if populations can be educated successfully it could mean that future generations are spared the grind of so-called 'sweatshop' employment (provided sufficient tertiary-sector work is created). Alternatively, wages may remain sufficiently low in some African and South Asian economies that it could make business sense to keep using human labour for years to come.

One thing is certain though. Any attempt to tackle perceived injustices linked with poor working conditions (in low-income countries) or jobs lost to offshoring (for example, in the USA) must also take into account trends in AI and robotics. Demands for higher-waged manufacturing work might just be the 'tipping point' that leads many TNCs to automate their operations further.

## Can global systems provide the technological fixes we need?

The UN has for many years led calls to make global systems more compatible with sustainable development principles and goals (see page 115). The section above dealing with climate change demonstrated how far reality continues to fall short of such aspirations. Indeed, current academic thinking in geography and other social sciences has begun to converge around a new concept called the Anthropocene which, in essence, takes a pragmatic view that major, irreversible environmental changes are now under way.

Central to much Anthropocene thinking is the sober realisation that technological fixes are the only realistic way of tackling pressing environmental issues such as climate change, biodiversity loss and plastic pollution in the Earth's oceans. High hopes are therefore attached to emerging technologies such as solar power, hydroponics and carbon capture and storage (CCS). Global systems may yet help humanity deliver these solutions.

- One result of global growth and development is the year-on-year growth in worldwide numbers of university graduates and new patented technologies. Provided universities and national governments are prepared to exchange data and ideas, rapid technological progress could deliver the solutions we need sooner rather than later.
- Once again, however, this is contingent on greater international collaboration. Unfortunately, recent heightened fears about cyber-security, data theft and fake news (see Chapter 3) lessen the chances of the USA, China and other world powers co-operating.

Pearson Edexcel

AQA

OCR

WJEC/Eduqas

### KEY TERMS

**Anthropocene** The idea that our planet has entered a new epoch where collective human actions are reshaping physical systems and the conditions of life on Earth as a whole.

**Technological fix** An innovation that can help humans overcome a pressing problem. Anthropological studies suggest that creativity is stimulated when societies are faced with environmental challenges, giving rise to new technologies that safeguard people's wellbeing.

# ③ Evaluating the issue

▶ *To what extent has a new era of deglobalisation begun?*

## Identifying possible contexts, criteria and evidence

This chapter's plenary section evaluates the concept of deglobalisation. If globalisation involves removing barriers to flows of money, people and ideas, then deglobalisation brings

the exact opposite. As this chapter has shown, people in many developed nations have begun distancing themselves from so-called globalists at the voting ballot box; they see globalisation as neither inevitable nor desirable. As a result, barriers to global flows are now rising in some parts of the world.

| Global flows and connections | Enquiry questions |
|---|---|
| Trade and investment flows | Has their value fallen, and have trade barriers appeared? What are the current trends for cross-border foreign investment, mergers and acquisitions? |
| International agreements | On balance, does the evidence suggest there is increasing global co-operation or a gradual retreat from multilateralism? What is the position of the main global superpowers? |
| International migration | Is the number of movers still on the rise globally? How many people are living outside of their country of birth compared with a few years ago, and what trends are predicted? |
| Flows of ideas and information | Has the shrinking-world effect peaked or will social networks continue to bring people and cultures together online, irrespective of where they are located in the real world? |

▲ **Table 6.2** Investigating deglobalisation: is a 'decoupling' of global connections really happening?

Deglobalisation (like globalisation) is a broad concept. Both its causes and effects have economic, social, cultural and political dimensions. These different strands can be evaluated systematically, as Table 6.2 shows.

It may be that while barriers to certain global flows are rising in some places, not all aspects of globalisation are rejected by those societies. Moreover, as this chapter's analysis of nationalism has shown, there is often a tension between what a country's *government* wants (and does) and the separate beliefs that *citizens* hold. In democracies, a majority of people may disapprove of the choices their politicians make. This has resulted in complex geographies of resistance to globalisation.

- The USA under President Trump withdrew from the UN Paris Agreement on climate change. However, many US states have decided independently to uphold their previous commitment to reducing carbon targets. Seattle City Council has a Climate Action Plan that still aims to make the city carbon-neutral by 2050. When media reporters say that 'the USA' is becoming more isolated, they are talking about the changing attitudes and actions of its country's government. But these attitudes are not shared by many of the USA's different states, cities and people.
- Similarly, many people in the UK voted for their country to leave the EU in the 2016

referendum. However, most of them will still want to travel abroad for holidays. They will continue to consume foods and TV shows from other countries. Also, UK businesses will still want to keep ties with the EU as close as possible, both for trade and as a source of workers.

Finally, we need to question what is meant by 'a new era' in the statement: 'To what extent has a new era of deglobalisation begun?' Perhaps a permanent change is under way; but then again, the Trump and Brexit era may yet turn out to be a temporary 'blip' in a longer globalisation timeline. Human and physical geographers alike use the idea of steady-state equilibrium when investigating temporal changes in systems.

- This is the concept of a system that changes but after a while returns to its original form (like a pendulum swinging slowly one way and then back again).
- Time will tell if the era of slower global systems growth and rising protectionism (which began in 2008 with the GFC) is a temporary pendulum swing or a permanent shift in direction. It will also become clear whether the nationalist course the USA adopted under President Trump in 2016 is here to stay or will reverse under a future leader.

# Evaluating changes in global trade and investment

Looking first at economic globalisation, there is plenty of evidence to suggest that barriers to global trade flows have recently increased in some places. Since the GFC of 2008–09, slower economic growth has led some governments to try to protect their industries in order to save jobs. As we have seen, record tariffs were imposed on US–China merchandise flows in 2018. Other examples of rising protectionism include the following.

- *Canada* – companies seeking to acquire a Canadian business must now ask for government approval under the Investment Canada Act. This involves passing a national security test and demonstrating the proposals will have a net benefit for Canada. Some deals have been blocked.
- *Australia* – the government has tightened rules on property purchases by overseas buyers and blocked bids for farmland by Chinese investors.

In 2016, it stopped the sale of a controlling state in the country's largest electricity network, Ausgrid, to a Chinese TNC.

## The reshoring phenomenon

According to Richard Ward, ex-chief executive of Lloyd's of London:

*'What people are waking up to is the interconnectedness of global trade – a single missing chip from Japan can shut down an (American) Ford factory on the other side of the world.'*

Page 106 explored the multiplying human and physical risks that businesses are exposed to when they develop extensive global production networks (GPNs). Some TNCs are now trying to adapt to these risks by building more resilient supply chains. This involves spreading their operations less thinly than before across fewer countries. Additionally, some US companies may be answering calls from President Trump to 'bring back American jobs' lost to global shift. Table 6.3 shows recent examples of the reshoring and nearshoring of TNC operations.

| Apple | ■ Apple's supply-chain problems were described in Chapter 3 (see page 89). In 2012, the company announced its intention to reshore parts of its operations, even though labour costs are higher back in the USA. The company has invested US$100 million in making some of its products at home again. |
|---|---|
| Other US TNCs | ■ Retail giant Walmart has committed itself to increased spending on goods sourced in the USA in future. Currently, many of the goods it needs are outsourced to companies in China and Vietnam. However, rising labour costs in China and supply-chain disruptions in Vietnam due to political protests have reduced the cost savings made through outsourcing.<br><br>■ General Electric has spent nearly US$1 billion re-establishing manufacturing at a facility it had all but abandoned in Kentucky; Otis has brought elevator production back from Mexico to South Carolina; Wham-O has brought Frisbee-making back from China to California. |
| UK food retailers | ■ Meat supply chains for UK supermarkets shortened after horsemeat was found in products labelled as beef in 2013. Complex food supply chains routed through France, Luxembourg, Cyprus, the Netherlands and Romania were to blame: at some point in the supply network, deliberate mislabelling had taken place. The UK press acted hysterically (British people do not, as a rule, view horses as a food source). Sales of supermarket meat plummeted.<br><br>■ Retailing TNCs woke up to the fact that their supply chains had become too complex to monitor satisfactorily; many have now increased their use of meat from local suppliers within the UK. |
| The aerospace industry | ■ Boeing and Airbus both shortened their supply chains when new aircraft models were recently introduced. The aim is to improve the resilience of their supply chains. Highly specialised parts design means there are very few back-up suppliers either firm can use if lengthy supply chains are disrupted (see Figure 6.16). 3D printing is also helping to shorten supply chains for this industry. |

▲ **Table 6.3** Recent examples of reshoring and nearshoring by TNCs

Reshoring is not the only way to manage risk of course. Alternatively, TNCs can:

- extend their GPNs further to include alternative back-up sources for goods and services, thereby building resilient supply chains
- introduce more rigorous checks and service-level agreements (see page 151); their own employees can be placed in the contractor's facility, to keep a closer eye on things.

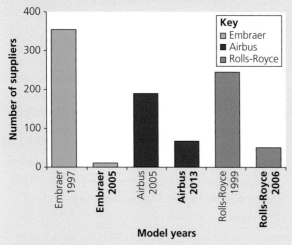

▲ **Figure 6.16** Reducing risk in the aircraft industry: the latest models (shown in bold) use far fewer different suppliers

## The changing political landscape of trade agreements

The UK's Brexit referendum result and the tough nationalist rhetoric of some prominent world leaders – including US President Trump, Russia's President Putin and Brazil's President Bolsonaro – can make it seem that global support for international co-operation on trade and other matters is waning. However, history may yet view these events and people as temporary disruptive elements in a bigger picture of expanding global partnerships and long-term growth.

- Large trading areas like the EU and South America (Mercosur) continue to prosper and may expand their membership further in the future. Brand new free-trade deals continue to be struck between many players, for example between the EU and Canada in 2017.
- The vast, new Pacific Rim pact (known as the Comprehensive and Progressive Agreement for Trans-Pacific Partnership, or CPTPP) has made progress despite the USA walking away from negotiations. It involves Japan and ten other countries: Malaysia, Vietnam, Singapore, Brunei, Australia, New Zealand, Canada, Mexico, Chile and Peru.
- Although EU sanctions led to reduced Russo-European trade flows after the Crimean crisis, Russia immediately entered into new agreements with India, Turkey and China (the latter is now the leading destination for Russian exports, doubling in value from US$20 to 40 billion between 2016 and 2018).
- There is no sign of declining inward investment into popular outsourcing destinations like Bangladesh or India. Broad trajectories for both world trade, GDP growth and international corporate mergers all headed upwards in 2017 and 2018, albeit at slower growth rates than those recorded in the early 2000s.

In summary, one emerging view of world trade is that 'business as usual' globalisation has continued, but without the USA in the driving seat.

# Evaluating changes in global migration

In some parts of the world, barriers to migration have risen. The UK voted to leave the EU in large part due to many British citizens' antipathy towards immigration. Many EU citizens have since left the UK because of uncertainty about the future (see Figure 6.17). Elsewhere, new nationalist movements are in ascendency throughout Europe and North America (see pages 176 and 177).

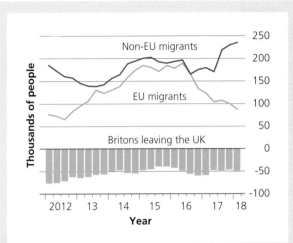

▲ **Figure 6.17** EU net migration into the UK fell in 2018 to its lowest level for many years. Uncertainty over the future has prompted many EU citizens to question whether they have a future living in the UK. *FT Graphic; Source: ONS*

However, the ageing population challenge faced by countries in these regions means long-term obstacles to immigration are potentially self-defeating (see page 186). In the future, we may see fewer rather than more barriers to movement. This view is supported by the Japanese government's recent move to soften migration and citizenship rules (see page 48).

In other parts of the world there is growing enthusiasm for free movement of people and visa-free travel. The African Union is taking steps to make movement easier for all 55 of its member states. South American countries have also agreed to make temporary residency rights easier to gain. Therefore, it is unclear whether global migration is rising or declining overall. Currently, a record number of 250 million people live outside the country they were born in and it seems unlikely that this number will fall any time soon.

## Evaluating changes in global data flows

Finally, consider global data and information flows. They play a vital role in economic globalisation because services and goods are increasingly traded internationally online. Data flows also contribute to cultural globalisation by sharing music, ideas, languages and other aspects of culture. You will know from your own experiences how essential the internet has become as an educational tool. Students across the world rely increasingly on formal global news sites like CNN and Al Jazeera alongside Wikipedia's 'collective commons' information store.

It is true that many states do not grant citizens unrestricted access to the internet, preferring instead to develop a 'splinternet' which is only partially integrated into the worldwide web. Around 40 world governments limit their citizens' freedom to access online information (violent or sexual imagery is censored in many countries; however, a 'dark web' also exists which is harder to control). But despite these and other restrictions, global data flows keep growing as more people acquire smartphones or other networked devices. The DHL Global Connectedness Index, like the KOF Globalisation Index (see page 16), triangulates multiple data sources in an attempt to quantify global flow trends. Figure 6.18 shows DHL's findings for 2018: cross-border data flows increased by 60 per cent between 2005 and 2015 and are expected to keep rising steeply as more people in emerging economies cross the digital divide and gain online access.

## Reaching an evidenced conclusion

There is little doubt that the 'golden age' of globalisation which lasted from the 1980s to the early 2000s has now ended. The GFC was the first sign that globalisation might have a 'reverse gear'. The abrupt decoupling of global connections in 2008 highlighted the amplified risks of interdependence. Wealth was annihilated on an unprecedented scale during the crisis. Since then, global trade growth has remained relatively depressed compared with

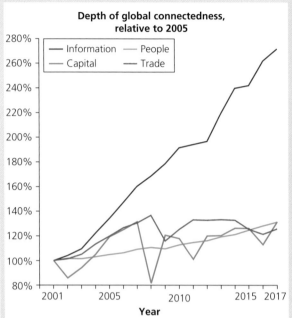

**Depth of global connectedness, relative to 2005**

▲ **Figure 6.18** The DHL Global Connectedness Index (2018 report) – which tracks trade, money (capital), people and information flows – suggests globalisation has not reversed

| 2013 | The EU expanded further to include Croatia; other Balkan states hope to join soon. |
| 2015 | The UN Paris Climate Change Conference was a success story for global governance in many people's view, despite the USA's later withdrawal. |
| 2015 | Chinese investment in Europe and the USA reached a record total of almost US$40 billion. |
| 2016 | The African Union announced plans for all its member states to seek visa-free travel. |
| 2016 | The number of international migrants reached a record 250 million. |
| 2017 | Facebook first registered 2 billion users, representing unprecedented human connectivity. |
| 2018 | The 11-member CPTPP trade agreement entered its final ratification phase. |

▲ **Table 6.4** Signs of the strengthening, not weakening, of global systems and interdependence

previous decades. This slowdown – when combined with resurgent nationalism and protectionism – may seem to suggest that globalisation has stalled or retreated. However, other global flow trends shown in Figure 6.18 – especially information flows data – indicate globalisation is far from being a spent force. Table 6.4 summarises some key facts that can also be used to refute the claim that a new era of deglobalisation has begun.

The concept of deglobalisation is useful insofar as it prompts us to analyse and evaluate how global systems are changing and evolving over time. However, the view that globalisation – in its totality – has somehow completely 'reversed' is unduly simplistic. It is more the case that global systems have entered a new phase characterised by changes in the relative importance of different flows.

- Some commodity flows and financial flows have reduced in size or are growing more slowly than in the past. The growth of protectionism means trade in merchandise is unlikely to accelerate any time soon.
- In contrast, data flows continue to balloon in size, along with worldwide social media use.
- Global migration flows will most likely continue to rise unless far more countries reinstate border controls. Remittance flows will most likely keep growing in line with migration.

The world is therefore still shrinking in many respects, despite uncertainty over future trends in cross-border physical trade flows.

Global systems are changing in another important way too: cultural and economic integration is less obviously driven by the USA since President Trump took office. Perhaps it makes more sense to argue that we are entering a 'de-Americanised' – rather than deglobalised – historical era.

- One view is that China – propelled by its Belt and Road Initiative (see page 66) – could assume the mantle of global leadership in place of the USA.

- In fact, the transfer of power to China pre-dates the arrival in the White House of President Trump. Cracks in the edifice of US hegemony date back to al-Qaeda's attack on the World Trade Center in 2001. One lasting effect of the GFC in 2008 was to discredit the neoliberal economic philosophy that policy-makers in Washington, DC, had for decades promoted. Their system approach suffered catastrophic failings during the GFC.

- It is perhaps no wonder that so many developing countries are happy to join with China's Belt and Road Initiative as an alternative to seeking help from the Washington-headquartered IMF and World Bank.

New geographies of globalisation are now forming, shaped increasingly by China and other emerging powers, along with a more aggressive Russian Federation. These countries are following different development paths from those previously walked by Europe and North America. As a result, it is hard to see how 'business as usual' US-led globalisation can ever be resumed. It is better perhaps to say we have moved from 'Globalisation 1.0' to 'Globalisation 2.0' rather than argue the party is over.

 **KEY TERMS**

**Steady-state equilibrium** A long-term balance is maintained, although there may be short-term changes in the system's state.

**Nearshoring** This involves avoiding distant outsourcing destinations and making use of companies in neighbour or near-neighbour states instead. This can reduce the risks and costs associated with longer-distance outsourcing.

# Chapter summary

✔ More governments are showing signs of listening to voters who are sceptical of globalisation; many citizens in developed countries now take the view that globalisation has unjustly benefited a global elite but not ordinary people. The result had been increased ballot-box support for nationalist parties and polices.

✔ Opposition to international migration, offshoring and trade bloc membership has increased in some countries, especially among social groups who believe they have not gained from globalisation and interdependence. This can be seen in the UK and USA in particular.

✔ World trade and financial flows have slowed since 2008 on account of the global financial crisis and the difficulties some emerging economies, notably China, have experienced while trying to maintain previously high growth rates. Other pressing and emerging challenges for global systems include: climate change; demographic issues (persisting high fertility in some countries and ageing populations in others); and the implications of new technologies for global employment.

✔ The concept of globalisation is complicated and consists of numerous processes and flows, not all of which have been paused or reversed. Although China's growth and global commodity trade have both slowed, global internet use continues to accelerate.

## Refresher questions

1 What is meant by the following geographical terms? Nationalism; protectionism; deglobalisation.

2 Outline the reasons why global economic growth has fallen in some past years before rising again.

3 Using examples, explain what is meant by 'reshoring'.

4 Using examples, outline reasons for the resurgence of nationalism in some developed countries.

5 Explain why global trade growth has remained relatively low since 2008.

6 What is meant by the following geographical terms? Demographic dividend; ageing population; premature deindustrialisation; Anthropocene.

7 Compare and contrast recent growth trends for trade flows, migration flows and data flows.

8 Outline recent examples of (i) new international trade agreements and (ii) policies to encourage migration.

## Discussion activities

1 In pairs, discuss the implications of Africa's population growing from 1 to 4 billion by 2100. Think about how this could affect the volume and pattern of different global flows (merchandise, investment and people).

2 In small groups, discuss the implications of AI for global systems. 'Robo-advisers' could lead to the loss of traditional high-street travel agents; driverless cars could leave taxi drivers unable to earn a living. What jobs, if any, do you think are immune to the rise of AI? Will AI lead to greater growth and development or increased injustice and inequality in the future?

3 In small groups, discuss the alternative possibilities that may exist to the current model of global capitalism. Could globalisation evolve in the future to embrace socialist (sharing) principles instead of neoliberalism and free markets?

4 Discuss the changes in technology that have taken place in your own lifetime. What technologies or apps do you currently rely on that used to be unavailable to people of the age you are now? How, if at all, do these technologies make you feel part of a global society or system?

5 As a whole-class exercise, discuss the following plenary questions.
   - Some people view the UK population's vote in 2016 to leave the EU as illogical, given the urgent need for greater international co-operation to tackle existential threats ranging from climate change and pandemics to terrorism. Others say it was the right thing to do. What's your view?
   - Could globalisation accelerate once again with the USA as the leading global power or is the long-term balance of power over global systems shifting permanently towards China? How much difference could a new US president make in the future?
   - Which contemporary challenge(s) do you see as being the greatest obstacle to sustained economic growth for the global community? Give reasons for your answer.

# FIELDWORK FOCUS

Globalisation, migration, sovereignty and the UK's changing relationship with the rest of Europe (see pages 175–180) are issues which have dominated politics and the news in recent years. There is plenty of scope to carry out a survey of social attitudes towards these and other contemporary issues. It might be interesting to devise a stratified sample, for example by focusing on different age groups, or populations living in two contrasting areas.

Alternatively, you might investigate other forms of resistance to globalisation, such as local sourcing movements, or opposition to the arrival of a TNC (such as McDonald's) in a town centre (some local people may be concerned that their home place is becoming a so-called clone town).

The technological changes outlined on pages 189–191 (including growth in online shopping and new developments in artificial intelligence) are beginning to impact on high streets throughout the UK. Your independent investigation could use a mixture of secondary and primary data to explore how far changes in retailing footfall in a particular place may be related to the local community's adoption of online shopping (which has been made possible by the growth of global information systems).

# Further reading

Castles, C. and Davidson, A. (2000) *Citizenship and Migration: Globalization and the Politics of Belonging*. London: Routledge.

Ghemawat, P. (2017) Globalization in the Age of Trump. *Harvard Business Review*. Available at: https://hbr.org/2017/07/globalization-in-the-age-of-trump.

IMF Data Mapper. Available at: www.imf.org/external/datamapper/NGDP_RPCH@WEO/OEMDC/ADVEC/WEOWORLD.

Institute of Fiscal Studies (2018) *10 Years On – Have We Recovered From the Financial Crisis?* Available at: www.ifs.org.uk/publications/13302.

MacKinnon, D. and Cumbers, A. (2018) *An Introduction to Economic Geography: Globalization, Uneven Development and Place*. London: Routledge.

# CHAPTER 7

# Study guides

 AQA A-level Geography: Global Systems

## Content guidance

The compulsory topic of Global Systems (Topics 3.2.1.1, 3.2.1.2, 3.2.1.3 and 3.2.1.6) is fully supported by this book. Global Governance (Topic 3.2.1.4) and The Global Commons (Topic 3.2.1.5) are supported by a separate title in Hodder's A-level Geography Topic Master series (shown on the inside back cover of this book). As part of the AQA assessment, students are expected to apply a wide range of geographical skills (3.2.1.7), including data manipulation and statistics, along with evaluative essay writing.

### Development terminology and case studies

The preferred terms for the AQA course are as follows:

- *Highly developed economies* (in this book, the terms 'high-income country', 'developed country' or 'advanced country' are often used instead).
- *Emerging major economies* (in this book, the term 'emerging country' is sometimes used instead).
- *Less-developed economies* (in this book, the terms 'developing country' or 'low-income country' are sometimes used instead).

Detailed case studies are required of *one* TNC and world trade in *one* food or manufacturing commodity.

| Sub-theme and content | Using this book |
|---|---|
| **3.2.1.1 Globalisation** | Chapter 1, pages 1–23 |
| This section introduces the key concepts and terminology required for the study of global systems. This includes recognition of the main flows of capital, labour, products, services and information, and also the global production networks of businesses (marketing, production, distribution and consumption). An overview is required of the factors responsible for globalisation. These are communications and transport technology, financial relationships and trade agreements. | Chapter 2, pages 41–61 |

| Sub-theme and content | Using this book |
|---|---|
| **3.2.1.2 Global systems**<br><br>There are economic, political, social and environmental forms of interdependence. Students should be familiar with the main issues which interdependence is linked with:<br><br>■ Unequal flows of people, money, ideas and technology within global systems can sometimes act to promote stability, growth and development but can also cause inequalities, conflicts and injustices for people and places.<br><br>■ Unequal power relations enable some states to drive global systems to their own advantage and to directly influence geopolitical events, while others are only able to respond or resist in a more constrained way. | Chapter 3, pages 78–85<br><br>Chapter 3, pages 100–111<br><br>Chapter 2, pages 62–77 |
| **3.2.1.3 International trade and access to markets**<br><br>This is a substantial section of the specification.<br><br>■ First, students must be familiar with the main features, trends and patterns of contemporary international trade and investment.<br><br>■ Second, there is a focus on the main trade relationships between highly developed economies, emerging economies and less-developed countries, e.g. in sub-Saharan Africa. The uneven impacts of trade and differential access to markets on countries of different levels of development should be explored.<br><br>■ The nature and role of TNCs should be studied. This includes their spatial organisation, production, linkages, trading and marketing patterns.<br><br>■ This must be accompanied by a detailed case study of *one* TNC and a case study of world trade in *one* food or manufacturing commodity.<br><br>■ Finally, students must analyse and assess ways in which trade and uneven access to markets impacts on people's lives in different societies. | Chapter 4, pages 112–118<br><br>Chapter 1, pages 24–40<br><br>Chapter 4, pages 128–143 |
| **3.2.1.6 Globalisation critique**<br><br>Here, the focus is the impacts of globalisation, including both benefits and costs. Possible benefits include growth, development, integration and stability.<br><br>Costs may include inequality, injustice, conflict and environmental impacts. | Chapter 2, pages 86–99<br><br>Chapter 4, pages 118–127<br><br>Chapter 5, pages 144–162<br><br>Chapter 6, pages 172–199 |
| **3.2.1.7 Quantitative and qualitative skills**<br><br>Students must ensure they can apply their full range of skills to the study of global systems. This includes familiarity with complex index and flow charts which can be used to illustrate trends and patterns in globalisation, development and inequality. | All chapters |

# AQA assessment guidance

Knowledge of global systems and flows is assessed as part of Paper 2 (7037/2). This examination is 2 hours and 30 minutes in duration and has a total mark allocation of 120. There are 36 marks allocated to the entirety of Global Systems and Global Governance. This consists of:

- a series of three short-answer questions, worth 16 marks in total, covering both Global Systems (this book) and Global Governance
- one 20-mark evaluative essay (in any year, the focus could be *either* Global Systems *or* Global Governance).

## AQA short-answer questions (up to 16 marks)

Your first question could be a knowledge-based short-answer task worth 4 marks and targeted at AO1 (assessment objective 1 – knowledge and understanding) using the command term 'explain'. High marks will be awarded to students who can write concise, detailed answers which link together a range of ideas, concepts or theories. As a general rule, try to ensure that every point is either developed or exemplified:

- A developed point takes the explanation a step further (for example, by adding extra detail of how a process operates).
- An exemplified point refers to a relatively detailed or real-world example in order to support the explanation with evidence.

Your second short-answer question may make use of a resource (map/diagram/table) and is targeted at AO3 (assessment objective 3 – quantitative, qualitative and fieldwork skills). It will ask you to analyse or extract meaningful information or evidence from the information provided. The question will most likely use the command terms 'analyse', 'compare' or 'assess'. The Analysis and interpretation questions included throughout this book are intended to help support the study skills you need to answer this type of question successfully.

Your third and final short-answer question will again make use of a figure but is now targeted mainly at AO2 (assessment objective 2 – application of knowledge and understanding). It will use a command phrase such as 'Analyse the figure and using your own knowledge…' This means you are expected to use the data only as a springboard to apply your own geographical ideas and information. For example, a 6-mark question could accompany a graph or chart showing growth over time in the size of global data flows: 'Using the figure and your own knowledge, assess the relative importance of global data flows as a cause of economic growth.' You can answer using your own knowledge rather than extracting detailed information from the resource.

## Global Systems evaluative essay writing

The 20-mark Global Systems essay will most likely use a command word or phrase such as 'to what extent', 'how far', 'assess the extent' or 'discuss'. The mark scheme will be weighted equally towards AO1 and AO2. For instance:

**'The economic growth and development caused by global flows is usually accompanied by rising inequality and injustice.' To what extent do you agree with this view?**

The box on the next page provides guidance on how to answer this type of question.

# Writing an evaluative essay about Global Systems

Every chapter of this book contains a section called 'Evaluating the issue'. These have been designed specifically to support the development of evaluative essay-writing skills. As you read each 'Evaluating the issue' section, pay particular attention to the following.

■ *Underlying assumptions and possible contexts are identified at the outset.* Think very carefully about what kinds of contrasting contexts you could choose to write about. Consider the essay title: 'The economic growth and development caused by global flows is usually accompanied by rising inequality and injustice. To what extent do you agree with this view?' In your answer, you need to establish what contexts and criteria to write about. Flows could include: people, ideas, culture, money, goods and much more. Injustices include: land grabs, worker exploitation, cultural erosion, environmental degradation etc.

■ *Extended writing can be structured carefully around different paragraphed themes, views, concepts or scales of analysis.* Often, essay questions will ask you to discuss or evaluate the 'role', 'significance', 'importance' or 'benefits' of something (in relation to different global system flows or impacts, for example). Consider this exam-style essay question: 'Assess the role of different global flows on the economic development process in different countries and locations.' Ask yourself: what different flows can be written about when responding? What is the time scale for these flows and development processes? Do all locations in a country experience the costs and benefits of migration flows or just particular hub

regions and settlements? Important questions such as these should be thought about at the planning stage of your essay and may help form an introduction.

Command words and phrases such as 'evaluate', 'to what extent' and 'discuss' require you to reach a final judgement. Draw on all the arguments and facts you have presented in the main body of the essay, weigh up your evidence and say whether – on balance – you agree or disagree with the question you were asked. To guide you, here are three simple rules.

■ *Never sit on the fence completely.* Essay titles are created purposely to generate a discussion which invites a final judgement following a debate. For example, the question: 'To what extent has globalisation benefited all places and societies?' Do not expect to receive a really high mark if you end your response with the sentence: 'So all in all, some places have benefited but others have not.'

■ *Equally, it is best to avoid extreme agreement or disagreement.* In particular, you should not begin your essay by dismissing one viewpoint entirely, for example: 'In my view, globalisation has greatly benefited the world, and this essay will explain all of the reasons why this is the case.' It is essential to consider a range of arguments or points of view.

■ *An 'agree, but...' or 'disagree, but...' judgement is usually the best position to take.* This is a mature viewpoint which demonstrates you are able to take a stand on an issue while remaining mindful of other views and perspectives.

## Synoptic geography

In addition to the three main AOs, some of your marks are awarded for 'synopticity'. The box below explains what this means.

# Thinking synoptically

Instead of focusing on one isolated topic, you are expected to draw together information and ideas from across your specification. You will be making connections between different 'domains' of knowledge, especially links between people and the environment (i.e. connections across human geography and physical geography). The section on nationalism in Chapter 6

is a good example of synoptic geography because of the important linkages between global systems and impacts on changing local places (especially in relation to migration); so too is the study of climate change refugees (see page 162) because it links together globalisation with carbon-cycling dynamics.

Throughout your course, take careful note of synoptic themes whenever they emerge in teaching, learning and reading. Examples of synoptic themes include: the impact of global investment by TNCs on local high streets; pressure on water cycle resources linked with the growth of emerging economies in global systems; links between global economic systems and global governance structures. Whenever you finish reading a chapter in this book, make a careful note of any synoptic themes that have emerged (they may have been highlighted or these could be linkages that you work out for yourself).

### AQA's synoptic assessment

Some 9-mark or 20-mark exam questions may require you to link together knowledge and ideas from different topics. These may appear in both your physical geography and human geography examination papers. For example, a Water and Carbon Cycles essay (paper 1) might ask you to think about ways in which physical systems could be affected by globalisation. A Changing Places essay (paper 2) might similarly require synoptic linkages to be established with Global Systems teaching and learning.

**How far do you agree that globalisation is a main cause of carbon and water cycle changes?**

**'The most important changes to local places are always a result of global flows of people, money, ideas or technology.' How far do you agree with this statement?**

For the second question, the mark scheme would include the following statement: 'This question requires links to be made across the specification specifically between Global Systems and content drawn from Changing Places.' One way to tackle this kind of potentially tricky question is to draw a mind map when planning your response. Draw two equally sized circles and fill these with relevant ideas, processes and contexts, trying to achieve the best balance you can between the two linked topics.

# Pearson Edexcel A-level Geography: (1) Globalisation and (2) Migration, Identity and Sovereignty

## Content guidance

The compulsory topic of Globalisation (Topic 3) is fully supported by this book. Material included in this book is useful too for the optional topic Migration, Identity and Sovereignty (Topic 8B).

### Development terminology and case studies

The preferred terms for the Pearson Edexcel course are as follows:

- *Developed country.* A country with very high human development (VHHD) (in this book, the terms 'advanced country', 'developed economy' or 'high-income country' are sometimes used instead).
- *Emerging country.* A country with high and medium human development (HMHD) – also, a recently emerging country (in this book, the term 'emerging economy' is sometimes used instead).
- *Developing country.* A country with low human development (LHD) – also, a poor country (in this book, the term 'low-income country' is sometimes used instead).

Study of detailed examples ('place contexts') is required in numerous places (see the specification for details of this, as denoted by the (🌍) symbol).

## Topic 3: Globalisation

The focus of Topic 3 is how globalisation and global interdependence have affected different places and people.

| Enquiry question and content | Using this book |
|---|---|
| **1 What are the causes of globalisation and why has it accelerated in recent decades?** | Chapter 1, pages 1–40 |
| | Chapter 2, pages 41–66 |
| First, this section looks at rapid developments in transport and communications over time leading to a shrinking world and time–space compression. Second, the focus shifts to the role of international organisations and national governments in creating frameworks for globalisation to operate. This includes a range of free-trade policies and measures to accelerate cross-border investment flows. Finally, students explore how globalisation has affected some places more than others. This includes describing different degrees of globalisation using the KOF Globalisation Index and explaining how TNCs contribute to the uneven spread of 'switched-on' and 'switched-off' places. There are physical and human reasons too why some countries remain largely isolated from globalisation. | |
| **Enquiry question and content** | **Using this book** |
| **2 What are the impacts of globalisation for countries, different groups of people and cultures and the physical environment?** | Chapter 2, pages 67–77 |
| | Chapter 3, pages 78–111 |
| This section looks at how global shift has brought a range of costs and benefits for different societies and environments. Important themes include the mixed blessings of globalisation for China and India, environmental issues for developing countries and issues to do with deindustrialisation in developed countries. Next, students look at the causes and consequences of megacity growth and key international migration flows. | Chapter 5, pages 144–150 |
| | Chapter 5, pages 157–163 |
| There are impacts for both host and source countries and regions. Finally, this section focuses on the emergence of a global culture as a result of globalisation. Important themes include global media and Westernisation; the spread of progressive attitudes, e.g. the success of the Paralympics; issues surrounding cultural erosion; and opposition to globalisation which is rooted in concern about cultural and environmental losses. | |
| **3 What are the consequences of globalisation for global development and the physical environment and how should different players respond to its challenges?** | Chapter 4, pages 112–124 |
| | Chapter 5, pages 163–172 |
| | Chapter 6, pages 173–199 |
| The final enquiry question is focused on the way globalisation is linked with development changes at different scales. This section begins by looking at different measures of development and recent trends in income inequality both globally and nationally. Students should also consider the link between economic development and environmental management. Next, the focus shifts to social and political tensions caused by rapid changes and the reaction to this including protectionism and isolation. The section finishes with a round-up of actions and strategies aiming to tackle ethical and environmental concerns about globalisation (such as local sourcing and fair trade). | |

## Topic 8B: Migration, Identity and Sovereignty

The focus of option 8B is the tension between globalisation and national sovereignty.

| Enquiry question and content | Using this book |
|---|---|
| **1 What are the impacts of globalisation on international migration?**<br><br>This section takes an in-depth look at migration both within and between countries, the reasons for movement and the cultural, political and economic issues for affected places. There is an opportunity to look at contemporary political events in the UK and USA which are linked with migration issues. | Chapter 3, pages 86–100<br><br>Chapter 4, pages 126–128<br><br>Chapter 5, pages 151–153 |
| **2 How are nation states defined and how have they evolved in a globalising world?**<br><br>This section begins with an overview of how nation states are defined and have developed over time. The difficulties of contested borders and claims for independence are explored. Next, students explore both historical and contemporary examples of nationalism, along with the persisting migration flows between former colonies and European countries. Finally, this section explores some economic issues relating to tax-haven states and the alternative economic development models followed by some South American countries. | Chapter 2, pages 41–45<br><br>Chapter 5, pages 157–160 |
| Enquiry question and content | Using this book |
| **3 What are the impacts of global organisations on managing global issues and conflicts?**<br><br>This section is focused on global governance. Students should understand how global organisations have developed over time, with special reference to (i) intergovernmental organisations responsible for global economic management (IMF, World Bank and WTO) and (ii) environmental management (climate change, biodiversity, oceans and Antarctica). | Chapter 2, pages 50–62<br><br>Chapter 5, pages 163–169 |
| **4 What are the threats to national sovereignty in a more globalised world?**<br><br>The last part of this topic explores new tensions between nationalism, cultural change and globalisation. Focuses include the management of multicultural societies, foreign ownership of different countries' businesses, the Westernisation of culture and foreign purchasing of property (e.g. Russian investment in London). This section concludes with a look at contemporary secession movements (Scotland) and the phenomenon of 'failed states'. | Chapter 2, pages 173–199<br><br>Chapter 6, pages 157–163 |

# Pearson Edexcel assessment guidance

Globalisation is assessed as part of 9GEO/02. This examination is 2 hours and 15 minutes in duration and has a total mark allocation of 105. There are 32 marks allocated to Globalisation/Superpowers. This is a combined assessment for Globalisation (Topic 3) and Superpowers (Topic 7). Note that Topic 7 is also supported by *Global Governance*, a separate title in Hodder's A-level Geography Topic Master series (shown on the inside back cover of this book).

The Globalisation/Superpowers assessment consists of:

- two short-answer questions (worth 4 marks each)
- two 12-mark evaluative 'mini-essays'.

Migration, Identity and Sovereignty is also assessed as part of 9GEO/02. There are 38 marks available, consisting of:

- a series of short-answer questions (worth 18 marks in total, including one 8-mark question)
- one 20-mark evaluative essay.

## Pearson Edexcel short-answer questions

Some questions will be knowledge-based tasks targeted at AO1 (assessment objective 1 – knowledge and understanding). These include:

- one or both of the 4-mark Globalisation questions
- the 8-mark Migration, Identity and Sovereignty question.

AO1 questions will use the command term 'explain'. High marks will be awarded to students who can write concise, detailed answers that incorporate a range of ideas, concepts or theories. As a general rule, try to ensure that every point is either developed or exemplified:

- A developed point takes the explanation a step further (for example, by adding extra detail of how a process operates).
- An exemplified point refers to a relatively detailed or real-world example in order to support the explanation with evidence.

Other short-answer questions may make use of a resource (map/diagram/table) and require use of a range of skills.

- Note that the Pearson Edexcel examination does *not* employ descriptive written AO3 (assessment objective 3) tasks such as 'describe the pattern shown in the figure' or 'analyse the trends shown in the figure'. However, you *could* be required to briefly complete a short skills-based numerical or graphical AO3 task. The specification includes a list of skills and techniques you are expected to be able to carry out, such as a Spearman's rank correlation, the calculation of an interquartile range or accurate plotting of data onto a chart or graph.
- In Migration, Identity and Sovereignty, expect a 6-mark short-answer question which makes use of a figure. It will be targeted in part at AO2 (assessment objective 2 – application of knowledge and understanding). It will most likely use the command phrase: 'Suggest reasons...' This means you are expected to use the data only as a springboard to apply your own geographical ideas and information. For example, a 6-mark question might ask: 'Suggest reasons for the differences in the proportion of migrants living in each country shown in the figure.' You can answer by applying your own knowledge and understanding in order to account for any patterns, trends or correlations.

## Evaluative essay writing

The two 12-mark Globalisation/Superpowers mini-essays and the 20-mark essay (Globalisation and Migration, Identity and Sovereignty) will use the command words 'assess' and 'evaluate' respectively. Both will have a mark scheme which is weighted heavily towards AO2. For instance:

**Assess the impacts of globalisation on different development gaps and disparities. (12 marks)**

**Evaluate the extent to which international organisations are the most important factor responsible for increased migration flows. (20 marks)**

The box on page 203 provides guidance on how to answer these questions.

## Synoptic geography

In addition to the three main AOs, some of your marks are awarded for 'synopticity'. The box on page 203 explains what this means.

### Pearson Edexcel's synoptic assessment

Synoptic exam questions are worth plenty of marks and you need to be well-prepared for them.

- In the Pearson Edexcel course, an entire examination paper is devoted to synopticity: Paper 3 (2 hours 15 minutes) is a synoptic 'decision-making' investigation. It consists of an extended series of data analysis, short-answer tasks and evaluative essays (based on a previously unseen resource booklet).
- As part of your Paper 3 answers, you will need to apply a range of knowledge from different topics you have learned about and make good analytical use of the previously unseen resource booklet (the Analysis and interpretation questions in this book have been carefully designed to help you). The context used in the resource booklet may well make use of themes drawn from Globalisation (Topic 3).

# OCR A-level Geography: Global Systems

## Content guidance

The topic of Global Systems is supported fully by this book. Students must choose from either Option A – Trade in the Contemporary World (Topic 2.2.1) or Option B – Global Migration (Topic 2.2.2). In each case, the detailed content is structured around three sub-themes. Note that the study of Global Governance (Topics 2.2.3 and 2.2.4) is supported by a separate title in Hodder's A-level Geography Topic Master series (shown on the inside back cover of this book).

## Development terminology and case studies

The preferred terms for the OCR course are shown on page 135. They are:

- *Advanced countries (ACs).* Countries which share a number of important economic development characteristics, including well-developed financial markets, high degrees of financial intermediation and diversified economic structures with rapidly growing service sectors. 'ACs' are as classified by the IMF (in this book, the terms 'high-income country' or 'developed country' are sometimes used instead).
- *Emerging and developing countries (EDCs).* Countries which neither share all the economic development characteristics required to be advanced nor are eligible for the Poverty Reduction and Growth Trust (PRGT) from the IMF. 'EDCs' are as classified by the IMF (in this book, the terms 'emerging economy' or 'emerging country' are sometimes used instead).
- *Low-income developing countries (LIDCs).* Countries which are eligible for the Poverty Reduction and Growth Trust (PRGT) from the IMF. 'LIDCs' are as classified by the IMF (in this book, the terms 'developing country' or 'low-income country' are sometimes used instead).

Detailed case studies are required of one AC, one EDC and one LIDC.

## Global Systems option A – Trade in the Contemporary World

| Enquiry question and content | Using this book |
|---|---|
| **1 What are the contemporary patterns of international trade?** This involves exploration of inter-regional and intra-regional trade patterns for merchandise, services and capital. These patterns can be related to global differences in economic development. Students should consider how trade can promote stability, growth and development but is linked with inequality, conflicts and injustices too. | Chapter 1, pages 1–17, 24–40 Chapter 3, pages 100–108 Chapter 4, pages 112–116, 133–143 |
| **2 Why has trade become increasingly complex?** This section looks at the factors that influence access to markets. They include communication and transport technology (allowing supply-chain growth), outsourcing by MNCs, the growth of trade blocs such as the EU, the growth of south–south trade between developing countries, services growth, labour mobility and the new international division of labour. Students must also research a case study of one EDC (this must include full consideration of its changing international trade pattern and the impacts and different forms of interdependence that trade gives rise to). | Chapter 1, pages 18–230 Chapter 2, pages 41–56, 61–67 Chapter 3, pages 78–94 Chapter 5, pages 1–17, 146–151 China can be used as a case study (pages 66–67, 117–119) |
| **3 What are the issues associated with unequal flows of international trade?** This final section involves two in-depth case studies. ■ Case study of one AC to show how core economies exert influence over global systems to their own advantage. This is illustrated through trade agreements, sustained economic growth and other opportunities. However, challenges must be recognised too, e.g. the US trade deficit with China. ■ Case study of one LIDC to show how peripheral countries lack influence over global systems and trading agreements. The balance of opportunities and challenges for such countries should be carefully analysed. | The USA can be used as a case study (pages 91–94, 173–183) The DRC can be used as a case study (pages 157–160) or Ghana (page 162) |

## Global Systems option B – Global Migration

| Enquiry question and content | Using this book |
|---|---|
| **1 What are the contemporary patterns of global migration?** | Chapter 1, pages 1–10 |
| This involves exploration of inter-regional and intra-regional migration patterns. These patterns can be related to global differences in economic development. Students should consider how migration can promote stability, growth and development but is linked with inequality, conflicts and injustices too. | Chapter 3, pages 78–81, 100–11<br><br>Chapter 4, pages 112–114, 125–128<br><br>Chapter 5, pages 151–157 |
| **2 Why has migration become increasingly complex?** | Chapter 1, pages 18–23 |
| This section looks at the factors that influence global migration patterns. They include communication and transport technology (allowing movement), the growth of new migration source and host areas in global systems, south–south migration flows between developing countries, conflict and changes in migration policies.<br><br>Students must also research a case study of one EDC (this must include full consideration of changing international migration patterns and flows, and the impacts and different forms of interdependence that migration gives rise to). | Chapter 2, pages 46–48, 57–59, 61–67<br><br>Chapter 3, pages 86–88, 95–99<br><br>China can be used as a case study (pages 66–67, 117–118) or Poland (pages 97–98) |
| **3 What are the issues associated with unequal flows of global migration?**<br><br>This final section involves two in-depth case studies.<br><br>■ Case study of one AC to show how core economies exert influence over global systems to their own advantage. This is illustrated through migration patterns and agreements, sustained economic growth and other opportunities associated with migration. However, challenges must be recognised too, e.g. EU border issues.<br><br>■ Case study of one LIDC to show how peripheral countries lack influence over global migration systems and agreements. The balance of opportunities and challenges for such countries should be carefully analysed. | The USA can be used as a case study (adapted from pages 91–94, 173–183, 191–199) or Poland (pages 97–98) or Qatar (pages 65–66)<br><br>The DRC can be used as a case study (pages 157–160) or Ghana (page 162) |

## OCR assessment guidance

Global Systems are assessed as part of Paper 2 (H481/02), in Section B (Global Connections). This examination is 1 hour and 30 minutes in duration and has a total mark allocation of 66. There are 17 marks allocated to Global Systems. In any year, this will consist of:

- *either* a series of three short-answer questions (worth 17 marks in total)
- *or* one 16-mark evaluative essay.

### Global Systems short- and medium-length questions

When the assessment is made up of short- and medium-length questions, it will include:

- low-tariff, point-marked questions (between 1 and 4 marks per question)
- one medium-length question (most likely 8 marks).

The short, low-tariff questions will be targeted jointly at AO3 (assessment objective 3) and AO2 (assessment objective 2). This means that you will be required to use geographical skills (AO3) to extract meaningful information or evidence from a resource, such as a photograph or chart; you will also be required to offer an explanation of this information using applied knowledge of global systems and flows (AO2).

- These questions are likely to include phrases such as 'Use the figure' or '...shown in the figure'.
- For example, a series of three low-tariff questions might accompany a map showing proportional flow arrows (representing either trade or migration volumes) linking different countries together. You might be expected to (i) analyse the patterns shown or make an appraisal of the way the data have been presented (AO3 tasks) and (ii) suggest reasons for variations in the sizes or lengths of the flows (AO2 tasks).
- The analysis and interpretation questions included in all chapters of this book are intended to support the study skills you need to answer these kinds of question successfully.

Finally, a medium-length question (around 8 marks), most likely using the command word 'explain', will be targeted at assessment objective 1 – knowledge and understanding. For example: 'With reference to a case study, explain how *either* trade *or* migration flows have led to increased interdependence between EDCs and other types of country.' High marks will be awarded to students who can write concise, detailed answers that incorporate a range of ideas, concepts or theories. As a general rule, try to ensure that every point is either developed or exemplified.

- A developed point takes the explanation a step further (for example, by adding extra detail of how a process operates).
- An exemplified point refers to a relatively detailed or real-world example in order to support the explanation with evidence.

## Evaluative essay writing

A 16-mark essay will most likely use a command word or phrase such as 'discuss', 'assess' or 'how far do you agree'. The mark scheme will be weighted equally towards AO1 and AO2. For instance:

**'The rich get richer and the poor get poorer because of international trade.' Discuss.**

**'The economic growth which global migration brings is always accompanied by rising inequality too.' Discuss.**

The box on page 203 provides guidance on how to answer this type of question.

## Synoptic geography

In addition to the three main AOs, some of your marks are awarded for 'synopticity'. The box on page 203 explains what this means.

### OCR's synoptic assessment

In the OCR course, part of Paper 3 (Geographical Debates H481/03) is devoted to synopticity. For this exam, you will have studied two optional topics chosen from: Climate Change, Disease Dilemmas, Exploring Oceans, Future of Food and Hazardous Earth. In Section B of Paper 3, you must answer two synoptic essays worth 12 marks each.

Each synoptic essay links together the chosen option with a topic from the core of the A-level course, such as Global Systems. Possible Paper 3 essay titles might therefore include:

**'Tectonic hazards usually have a negligible impact on global system flows.' For *either* global trade flows *or* global migration flows, how far do you agree with the statement?**

**Assess how the use of oceans is affected by *either* the global system of trade *or* the global system of migration.**

One way to tackle these kinds of questions would be to draw a mind map to help plan your response. Draw two equally sized circles and fill these with relevant ideas, processes and contexts, trying to achieve the best balance you can between the two linked topics. The mark scheme requires that your answer includes: 'clear and explicit attempts to make appropriate synoptic links between content from different parts of the course of study'.

# ④ WJEC and Eduqas A-level Geography: Processes and Patterns of Global Migration

## Content guidance

Both WJEC and Eduqas students must study the compulsory topic of Processes and Patterns of Global Migration (WJEC Topic 3.2.1–3.2.5; Eduqas Topic 2.2.1–2.2.5) which is supported fully by this book. Note that the study of Global Governance of Earth's Oceans is supported by a separate title in Hodder's A-level Geography Topic Master series (shown on the inside back cover of this book). Additionally, the specification-matching grid inside the front cover indicates where some book content supports optional *Economic Growth and Challenges* topics.

### Development terminology and case studies

Preferred terms for the WJEC and Eduqas courses are as follows:

- *Developed economies* (in this book, the terms 'high-income country', 'developed country' or 'advanced country' are often used instead).
- *Emerging economies* (in this book, the term 'emerging country' is sometimes used instead).
- *Developing economies* (in this book, the terms 'developing country' or 'low-income country' are sometimes used instead).

Detailed case studies are not required, though the use of illustrative examples is expected.

| Enquiry question and content | Using this book |
|---|---|
| **1 Globalisation, migration and a shrinking world** | Chapter 1, pages 1–40 |
| This section provides an overview of global systems and flows. It also includes the classification of migrants and the factors responsible for the 'shrinking-world' effect (transport and communications technology). | |
| **2 Causes of international economic migration** | Chapter 2, pages 41–67 |
| The focus here is factors driving international out-migration, including poverty and other injustices. Important drivers of migration include diaspora communities, international organisations (Commonwealth and EU) and the disproportionate influence of superpower states (as global migration hubs). | Chapter 3, pages 98–100 |

| Enquiry question and content | Using this book |
|---|---|
| **3 Consequences and management of international economic migration**<br><br>This section assesses the impacts of migration for places. This includes looking at the way global flows: can increase inequality or promote growth and stability; can play a role in fostering the interdependency of different countries and societies; may require careful management. | Chapter 3, pages 68–88, 101–111<br><br>Chapter 4, pages 133–143<br><br>Chapter 5, pages 163–172<br><br>Chapter 6, pages 173–199 |
| **4 Causes, consequences and management of refugee movements**<br><br>The focus here is the forced movement of refugees and Internally Displaced People (IDPS). The causes and consequences should be understood, including land grabs and other injustices. Students should also know about refugee management at global, national and local scales, and limitations of management (e.g. in conflict zones and borders in remote areas). | Chapter 2, pages 57–59<br><br>Chapter 5, pages 151–160 |
| **5 Causes, consequences and management of rural–urban migration in developing countries**<br><br>This final section deals with push and pull factors for rural–urban migration. These factors should be examined in a global systems context by looking at the way powerful global forces push people from the land and attract them to cities. Finally, students should take a brief look at management issues and strategies for growing urban areas of the developing world. | Chapter 4, pages 126–128<br><br>Chapter 5, pages 151–160 |

# WJEC and Eduqas assessment guidance

Global Systems is assessed as part of the following:

- *WJEC Unit 3.* This examination is 2 hours in duration and has a total mark allocation of 96. There are 35 marks allocated for the combined assessment of Processes and Patterns of Global Migration and Global Governance of Earth's Oceans, indicating that you should spend around 45 minutes answering. The 35 marks consist of:
  - two structured short-answer questions – one each on Processes and Patterns of Global Migration and Global Governance of Earth's Oceans (and together worth 17 marks)
  - one 18-mark evaluative essay (from a choice of two – one each on Processes and Patterns of Global Migration and Global Governance of Earth's Oceans).
- *Eduqas Component 2.* This examination is 2 hours in duration and has a total mark allocation of 110. There are 40 marks allocated for Global Systems, indicating that you should spend around 45 minutes answering. The 40 marks consist of:
  - two structured short-answer questions – one each on Processes and Patterns of Global Migration and Global Governance of Earth's Oceans (together worth 20 marks)
  - one 20-mark evaluative essay (from a choice of two – both are synoptic essays which draw equally on content from Processes and Patterns of Global Migration and also Global Governance of Earth's Oceans).

Both courses use broadly similar assessment models and these are dealt with jointly below.

## Short-answer questions

### WJEC

Short-answer questions 1 and 2 on your examination paper include several different types of short-answer question, usually following on from a figure (a map, chart or other resource).

Part (a) of one question – *but not the other* – will usually be targeted at AO3 (assessment objective 3) and is worth 3 marks. This means that you will be required to use geographical skills (AO3) to analyse or extract meaningful information or evidence from the figure. These questions will most likely use the command words 'describe', 'analyse' or 'compare'.

In part (b) of one of the questions, worth 5 marks, you may be asked to apply your knowledge and understanding of Global Systems in an unexpected way. This is called an applied knowledge task; it is targeted at AO2 (assessment objective 2). For example, you could be asked the question: 'Suggest reasons why the value of global remittances varies from year to year.' To score full marks, you must (i) apply geographical knowledge and understanding to this new context and (ii) establish very clear connections between the question that is being asked and the stimulus material (in this case, a graph showing the changing value of global remittances).

Your remaining short-answer questions will usually be purely knowledge-based, targeted at AO1 (assessment objective 1). They will be worth 4 or 5 marks and most likely use the command words 'explain', 'describe' or 'outline'. For example: 'Explain two reasons for high rates of rural–urban migration in emerging economies.' High marks will be awarded to students who can write concise, detailed answers which incorporate and link together a range of geographical ideas, concepts or theories.

## Eduqas

Short-answer questions 1 and 2 on your examination paper will be linked to figures (maps, charts, tables or photographs).

Part (a) of question 1 and part (a) of question 2 will always be targeted at AO3 (assessment objective 3). This means that you must use geographical skills (AO3) to analyse or extract meaningful information or evidence from the figure. These questions will most likely use the command words 'describe', 'analyse' or 'compare'.

In part (b) of one of these questions, you may be asked to suggest possible reasons that could explain the information shown in the figure. This question will usually be worth 5 marks.

- This is called an applied knowledge task; it is targeted at AO2 (assessment objective 2) and will most likely use the command word 'suggest'. An applied knowledge task will always include the instruction: 'Use the figure'.
- For example, two short questions could accompany a graph showing the changing value of global remittances in recent years. The opening part (a) question could be: 'Analyse the changes shown in the figure' (an AO3 task). The part (b) AO2 question that follows might ask: 'Suggest reasons why the value of global remittances varies from year to year.' To score full marks, you must (i) apply geographical knowledge and understanding to this new context and (ii) establish very clear connections between the question asked and the stimulus material.

The other part (b) question will usually be purely knowledge-based, targeted at AO1 (assessment objective 1), and worth 5 marks. For example: 'Outline reasons why developed countries usually experience high rates of rural–urban in-migration.' High marks will be awarded to students who can write concise, detailed answers which incorporate and link together a range of geographical ideas, concepts or theories.

## Evaluative essay writing

### WJEC

You are given a choice of two 18-mark (10 marks AO1, 8 marks AO2) essays to write (*either* question 3 *or* question 4). These essays will most likely use the command words 'assess' or 'examine'. For instance:

**Examine recent changes in the global pattern of migration.**

**Assess how far the economic growth and development caused by migration is usually accompanied by rising inequality and injustice.**

The box on page 203 provides guidance on how to answer these questions.

### Eduqas

You are given a choice of two 20-mark (10 marks AO1, 10 marks AO2) essays to write (*either* question 3 or question 4). These essays will most likely use the command words and phrases 'discuss', 'evaluate' or 'to what extent'. These essays draw equally on *both* global migration *and* ocean governance knowledge (the latter of these topics is not included in this book). For instance:

**'Barriers to global flows and movements are growing everywhere.' Discuss this statement (referring to both migration and ocean governance).**

**'The economic growth and development caused by global flows is usually accompanied by rising inequality and injustice.' Discuss this statement (referring to both migration and ocean governance).**

The box on page 203 provides guidance on how to answer these questions.

## Synoptic geography

In addition to the three main AOs, some of your marks are awarded for 'synopticity'. The box on page 203 explains what this means.

### The WJEC and Eduqas synoptic assessment

In the WJEC course, part of Unit 3 is devoted to synopticity while for Eduqas a similar assessment appears in Component 2. In both cases, synopticity is examined using an assessment called '21st Century Challenges'. This synoptic exercise consists of a linked series of four figures (maps, charts or photographs) with a choice of two accompanying essay questions. The WJEC question has a maximum mark of 26; for Eduqas it is 30. An example of a possible question is:

**'Physical processes can cause place identity to change rapidly whereas human activity always brings slower changes.' Discuss this statement.**

As part of your answer, you will need to apply a range of knowledge from different topics and make good analytical use of the previously unseen resources in order to gain AO3 credit (the Analysis and interpretation questions in this book have been carefully designed to help you in this respect). The topic of Processes and Patterns of Global Migration is highly relevant to the title shown above.

- Migration flows operate over varied time scales. It is possible for large numbers of people to arrive in a local places over a very short time (think of refugee movements).
- This essay therefore allows you to make varied arguments using knowledge of migration, along with ideas drawn from other different parts of the A-level specification.

# Index

# Acknowledgements

**p.9 Figure 1.6** Cross-border capital flows 1990-2017 from THE NEW DYNAMICS OF FINANCIAL GLOBALIZATION, AUGUST 2017 Copyright © McKinsey & Company. Reprinted with permission; **p.14 Figure 1.10** © Financial Times; **p.30 Figure 1.18** © Financial Times; **p.36 Figure 1.23** © Financial Times; **p.50 Figure 2.7** Vincenzo Cosenza vincos.it from Alexa/SimilarWeb covered under CC-BY-NC; **p.64 Figure 2.15** © Financial Times; **p.67 Figure 2.17** © Financial Times; **p.92 Figure 3.13** © Financial Times; **p.117 Figure 4.6** © Financial Times; **p.137 Figure 4.23** © Financial Times; **p.138 Figure 4.24** © Financial Times; **p.139 Figure 4.26** Lakner and Milanović 'elephant chart', Dr. Branko Milanović. Reprinted with permission; **p.155 Figure 5.10** Changes in the number of years a skilled service worker needs to work to be able to buy a 60m² flat from UBS Global Real Estate Bubble Index © UBS 2017. Reprinted with permission; **p.176 Figure 6.3** © Financial Times; **p.180 Figure 6.7** McKinsey Global Institute report, 2018: Poorer Than Their Parents? From McKinsey Global Institute report, 2018 © McKinsey & Company . Reprinted with permission; **p.180 Figure 6.8** Chart based on data from Eurobarometer (http://ec.europa.eu/public_opinion/archives/eb/eb69/eb69_globalisation_en.pdf); **p.184 Figure 6.11** Chart based on data from Institute of Fiscal Studies. Reprinted with permission; **p.186 Figure 6.13** © Financial Times; **p.188 Figure 6.14 (a) and (b)** © Financial Times; **p.195 Figure 6.17** © Financial Times; **p.196 Figure 6.18** Steven A. Altman, Pankaj Ghemawat, and Phillip Bastian, "DHL Global Connectedness Index 2018: The State of Globalization in a Fragile World," Deutsche Post DHL Group, 2019

This book contains excerpts of material and some illustrations that originally formed part of articles in *Geography Review* magazine or were included in the *Hodder book IB Diploma: Global Interactions* (2018).

**Photo credits**

**p.6** *t* Financial Times / FT.com. June 2016. Used under licence from the Financial Times. All Rights Reserved; *b* Gage Skidmore/https://commons.wikimedia.org/wiki/File:Donald_Trump_(25953705015).jpg/https://creativecommons.org/licenses/by-sa/2.0/deed.en[18Jan2019]; **p.26** Shutterstock / gnoparus; **p.33** *l* © Simon Oakes; *r* Public domain. https://commons.wikimedia.org/wiki/File:Michelle-obama-bringbackourgirls.jpg; **p.34** © Simon Oakes; **p.42** © Simon Oakes; **p.43** © Simon Oakes; **p.47** © ITAR-TASS News Agency / Alamy Stock Photo; **p.51** Shutterstock / Kristi Blokhin; **p.54** http://en.kremlin.ru/events/president/news/55515/photos/50064/https://creativecommons.org/licenses/by/4.0/[25Jan2019]; **p.60** ©Aleksandar Todorovic - stock.adobe.com; **p.66** © Simon Oakes; **p.73l** © Simon Oakes; **p.73r** © Stacy Walsh Rosenstock / Alamy Stock Photo; **p.70** VCG/VCG via Getty Images; **p.84** © Agencja Fotograficzna Caro / Alamy Stock Photo; **p.99** © Simon Oakes; **p.102** © Simon Oakes; **p.103** milesbeforeisleep/123RF.com; **p.104** © jryanc10 - stock.adobe.com; **p.107** © Martyn Evans / Alamy Stock Photo; **p.115** yavuzsariyildiz - stock.adobe.com; **p.124** © epa european pressphoto agency b.v. / Alamy Stock Photo; **p.126** © siempreverde22 - stock.adobe.com; **p.129** © Ajay Bhaskar/123RF.com; **p.130** © Ruslan Ivantsov - stock.adobe.com; **p.132** ©ah_fotobox - stock.adobe.com; **p.145** © David Hartley/REX/Shutterstock; **p.146** © Jake Lyell / Water Aid / Alamy Stock Photo. **p.148** © J Marshall - Tribaleye Images / Alamy; **p.151** © Tansh / Alamy Stock Photo; **p.157** © Jan Kranendonk/123RF.com; **p.158** *t* © Polyp.org.uk; *b* © Simon Oakes; **p.162** © Marlenenapoli/https://commons.wikimedia.org/wiki/File:Agbogbloshie.JPG/https://creativecommons.org/publicdomain/zero/1.0/deed.en; **p.167** © panya99 - stock.adobe.com; **p.169** © Kristian Buus; **p.175** © Andre Larsson/Nurphoto/REX/Shutterstock